METROLOGY AND DIAGNOSTIC TECHNIQUES FOR NANOELECTRONICS

METROLOGY AND DIAGNOSTIC TECHNIQUES FOR NANOELECTRONICS

Editor

Mudiwa Afolayan

scitus
academics

Metrology and Diagnostic Techniques for Nanoelectronics
Edited by **Mudiwa Afolayan**

ISBN: 978-1-68117-719-9
Library of Congress Control Number: 2016958818

SCITUS Academics LLC,
616, Corporate Way, Suite 2, 4766,
Valley Cottage, NY 10989

www.scitusacademics.com

Notice

Printed in United States of America on Acid Free Paper

PREFACE

Metrology is the science of measurements, and nanometrology is that part of metrology that relates to measurements at the nanoscale. Many governments and regulatory bodies worldwide have existing nanotechnology policies and are taking the preliminary steps towards nanometrology strategies, for example in support of pre-normative R&D and standardization work. The applications and economic impact of thin films for nanotechnological products are large, and extend over a wide range of industry sectors. As buried layers, thin films are core to the performance of state-of-art microelectronics and magnetic data storage devices.

Metrology and Diagnostic Techniques for Nanoelectronics highlights state of art developments in devoted to advancements in all the metrological aspects related to nanoscience and nanotechnology, critical for continuing technology scaling and product innovation. It offers an inclusive overview of engineering metrology and how it relates to micro and nanotechnology research and manufacturing. By combining established knowledge with the latest advances from the field, it presents a comprehensive single volume, written by renowned experts in the respective fields that can be used for professionals, researcher, practitioners and students. It is a complete guide for metrology and diagnostic techniques essential for process technology, electronics packaging, and product development and debugging.

The semiconductor industry faces significant challenges to continue increasing performance and functionality of information processing. New and improved metrology and characterization is required to support these advances in density and functionality. This book brings together scientists and engineers interested in all aspects of the characterization technology needed for nanoelectronic materials and device research, development, and manufacturing.

TABLE OF CONTENTS

1

BROADBAND MICROWAVE-BASED METROLOGY PLATFORM DEVELOPMENT: DEMONSTRATION OF IN-SITU FAILURE MODE DIAGNOSTIC CAPABILITIES FOR INTEGRATED CIRCUIT RELIABILITY ANALYSES

Lin You,[a] Chukwudi A. Okoro,[a] Jung-Joon Ahn,[a] Joseph Kopanski,[a] Rhonda R. Franklin,[b] and Yaw S. Obeng[a]

[a]Semiconductor and Dimensional Metrology Division, Physical Measurement Laboratory, National Institute of Standards and Technology, Gaithersburg, Maryland 20899, USA

[b]Department of Electrical and Computer Engineering, University of Minnesota, Minneapolis, Minnesota 55455, USA

ABSTRACT

In this paper, we discuss the use of broadband high frequency electromagnetic waves (RF) to non-destructively identify, classify and characterize performance-limiting defects in emerging nanoelectronic devices. As an illustration, the impact of thermal cycling on the RF signal characteristics (insertion loss (S_{21}) and return loss (S_{11})) is used to infer thermo-mechanical stress-induced defects in metal interconnects. The inferred defects are supported with physical analytical data where possible.

In this paper, we discuss the use of broadband radio frequency (RF) to non-destructively identify and characterize performance-limiting defects in emerging nanoelectronic devices. To do this, we reinterpret previously published (mostly from reference 1), and some new data, to demonstrate the utility of RF phase data in failure mode analysis, to detect structural changes in devices, without resorting to the traditional destructive physical analysis.

Emerging nanoelectronics are rapidly adapting three dimensional integrated circuits (3D-ICs), enabled by through-silicon vias (TSV), as a strategy to increase performance and functional diversification. However, the introduction of such 3D-ICs to the market has been hindered by reliability challenges such as stress related failures. Specifically, the stress buildup due to mismatch in the coefficient of thermal expansion (CTE) of the materials of construction results in the generation of defects such as cracks, voids, delamination, plastic deformation, substrate warping and buckling.[1] These types of damage ultimately lead to open or short circuits in the devices, resulting in catastrophic failure. While metrology for accurate quantification of the impact of stress evolution in interconnect structures is needed, further insights are required to characterize and understand the physical damage stemming from the stress buildup. Microwave (RF)-based metrology tools offer several advantages over the traditional techniques and are uniquely suitable for studying the buried structures and interfaces inherent in such nanoelectronic devices. For example, changes in RF insertion losses offer early prognostics of the onset of failure in ball grid arrays (BGA) subjected to accelerated temperature cycling tests.[2,3] RF measurements have also been used to specifically show that leakage conductance increased with the degree of voiding damage in Al and AlCu metal interconnects lines in integrated circuits.[3] As well, such high frequency measurements have been used to monitor TSV interconnect performance.[1]

When an electromagnetic wave propagating in a transmission line encounters a discontinuity, parts of the wave are either scattered, transmitted or reflected. The resulting signal losses, due to reflection and transmission, can be explained by changes in the resistive, inductive and/or capacitive properties of the line. Detailed study of the transmitted and reflected signal losses, therefore, can provide insights into physical changes in the device under test (DUT), since RF based measurements interrogate the entire device under test. Thus, intuitively, broadband RF interaction with devices potentially affords the diagnostics to enable in-situ deduction of failure modes and mechanisms without the need for physical failure mode analyses (FMA).

SAMPLE AND EXPERIMENT DESCRIPTION

Samples

Two types of interconnects will be discussed in this paper: the traditional planar AlSiCu-based local interconnects and the emerging TSV global interconnects to enable three-dimensional integration (3D-ICs).

For the planar local interconnects, we used a dedicated multi-metal level test structure, manufactured with the TSMC 0.35 μm process flow. The four metal levels, denoted as M1 (deepest) to M4 (closest to chip-air interface), respectively, are as depicted schematically in Figure 1.[c] In this representation, the top metal lines arrays (M4 and M3) are exposed to air, the middle metal line array (M2) is located about 800 nm beneath the bulk surface, while the bottom metal line array (M1) are located about 2000 nm below the dielectric surface (about 1000 nm below the M2) respectively.

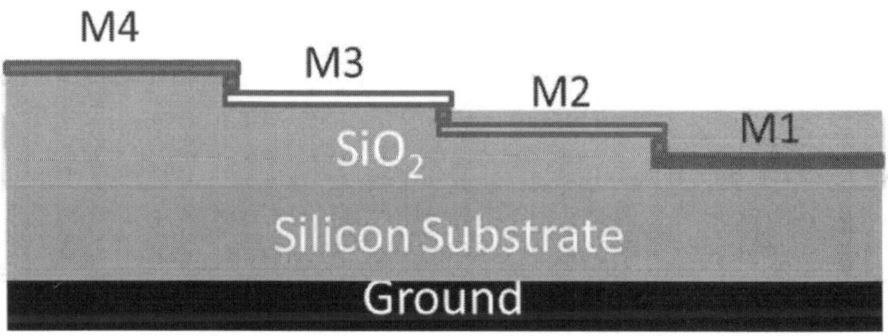

Figure 1. Schematic Plan view of 4-level BEOL test structure: M3 and M4 are exposed, while M2 and M1 lines are encapsulated in SiO_2.

For studying prototypical 3D-global interconnects, a two-level stacked-die mechanical test vehicle provided by SEMATECH (Albany, NY) was used. The TSVs were nominally 5.5 μm in diameter and 50 μm deep and were located in the top chip. The TSVs are isolated from the bulk silicon substrate by an isolation liner comprised of an insulating dielectric layer (0.5 μm thick SiO_2), barrier layer (0.025 μm thick TaN) respectively. The bottom and top chip wafers were bonded together with benzocyclobutene (BCB). Figure 2(a), shows the layout of the ground-signal-ground (GSG) configured test structure used in this work.

Figure 2(b) shows its cross-sectional scanning electron microscopy (SEM) image. The components of the two bonded wafer stack are clearly delineated in later figure. During the RF measurement, a ground-signal-ground (GSG) test probe was placed on the GSG bond pads at both ends of the test structure. For the proof of principle RF reliability study, thirty-four samples, chosen randomly across a single 300 mm wafer, were used.

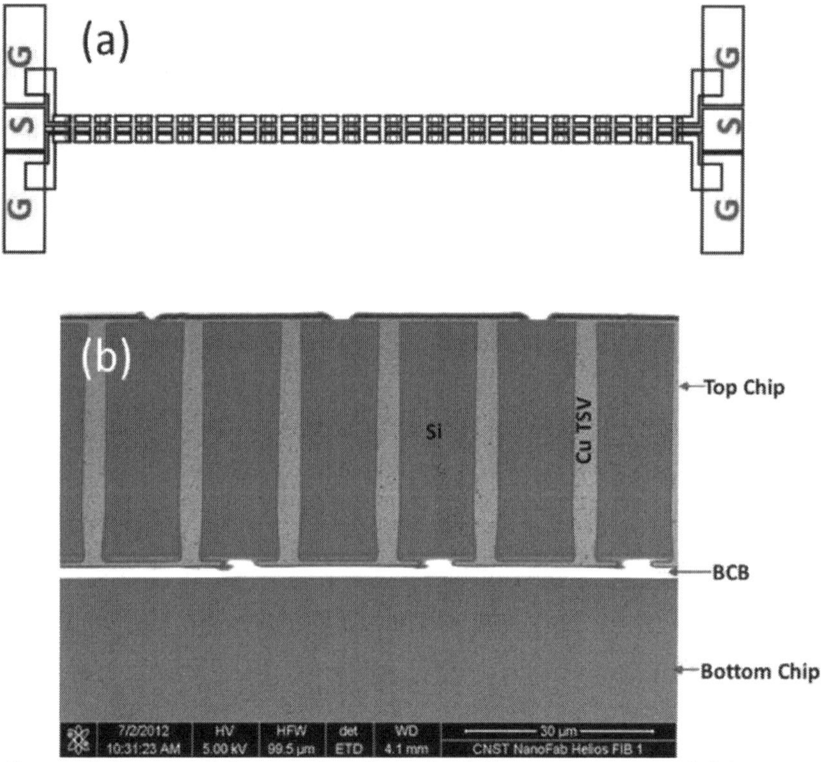

Figure 2. Image of the two-level stacked die test structure (a) Layout of the structure, having a GSG configuration (b) SEM cross-sectional image of the stacked chip.

Experimental

Thermal cycling was used to perturb and induce defects in the devices studied as described elsewhere.1 In brief, the samples were heated from 30°C to 150°C and then cooled to 30°C in laboratory ambient or under a nitrogen blanket. The insertion loss measurements, described elsewhere,1 were first measured on the devices under test (DUT) in the "as-received" state, prior to thermal cycling, to determine their baseline

RF signal characteristics. The DUTs were re-measured every 500 thermal cycles, for up to a maximum of 2000 thermal cycles, on a PNA N5230A (Agilent, Santa Clara, CA).[c, 1]

Scanning Microwave Microscopy (SMM) was used in studying the reflectance loss (S11) of the planar AlSiCu-based interconnect.4 The SMM implementation as discussed here, applies a high frequency signal through a 50 ohm transmission line terminated with a scanning probe microscope tip to the sample.4,5 The SMM generates spatially localized images from the reflected RF (S11) signal amplitude and phase, usually at resonance (around 7 GHz for the set-up used in these studies). The signal from the tip, depending on the skin depth, can reach relatively deep (on the order of tens of microns) into sample and interacts with the buried structures and interfaces. This interaction translates into the changes of the measured reflection coefficient S_{11}. As discussed below, we have demonstrated the subsurface imaging capabilities of SMM on buried integrated circuit metallization. To distinguish from 2-port RF scattering data presented later, scattering parameter data from the SMM method will be denoted as $S11_{SMM}$.

RESULTS AND DISCUSSION

During fabrication, and in-service usage, microelectronic devices are subjected to constant fluctuations in temperature. Such thermal cycling results in the buildup of stress in the TSVs.6 This occurs due to their confinement and the mismatch in the coefficient of thermal expansion (CTE) of Cu (16.7 ppm/°C), with respect to the surrounding Si matrix (2.3 ppm /°C), thus leading to the formation and growth of defects such as voids, and possible delamination at interfaces.7 The defects act as RF signal "scattering centers" resulting in the increase in the insertion loss. Thus, proper interpretation of the RF signal loss traces could help pinpoint or classify defect types. In this section, we demonstrate the inherent in-situ FMA capabilities of the RF-sample interaction by identifying and classifying the defect types suggested from the interpretation of the signal loss traces. Where possible, we support the defect classification with the results of physical analyses based on focused ion beam (FIB) based scanning electron microscopy (SEM) images.

Figure 3 compares the $S11_{SMM}$ amplitude and optical micrographs of the planar local interconnects (described in Figure 1) samples "as-received" ("a" and "c") to the same area after 1000 thermal cycles ("b" and "d"). The reflective losses were evaluated at 7 GHz on the scanning microwave microscope. For the M3 line arrays, the data show increased reflective signals (relative to the glass dielectric background) on the exposed metal lines after 1000 cycles as shown in the lower box of Figures 3a and 3b. The AFM topographical data, collected simultaneously with the $S11_{SMM}$ data, also showed enhanced roughness on the exposed M3 after the 1000 thermal cycles. Furthermore, detailed inspections of the optical micrographs, i.e., comparisons of Figures 3c and 3d, revealed extensive corrosion of the exposed M3 metal line array after the thermal cycling, when compared to the unexposed and insulated M2 line array. Thus, we attribute the increased reflective RF, and increased roughness to the presence of dielectric corrosion byproducts on the M3 after thermal cycling. The corrosion of the metal lines is possibly due to traces of oxygen and water vapor in the nitrogen ambient used during the thermal cycling.8

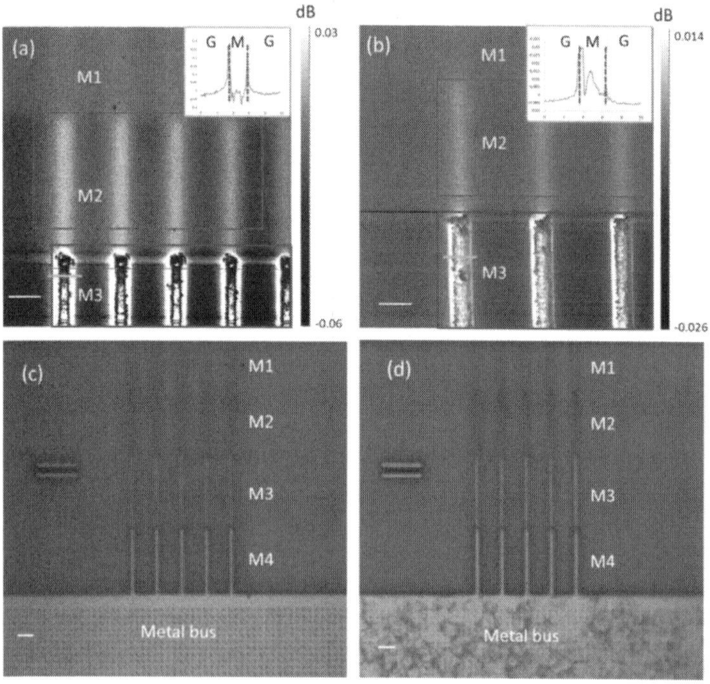

Figure 3. (a) $S11_{SMM}$ amplitude map of 'as-received" (b) $S11_{SMM}$amplitude map of representative features after 1000 thermal cycles at 7 GHz. Insets are line profiles across single M3 traces, to show the relative RF reflectance off the metal and the surrounding dielectric (from the $S11_{SMM}$ maps, ."M" denotes metal while "G" stands for glass).

(c) and (d) are optical images before and after 1000 thermal cycles, respectively. The white scale bars in each image represents 5 μm.

Transmission data is also obtained based on a two-port measurement on a PNA. The through-reflect-line (TRL) calibration method is used to remove cables and connectors in order to establish the reference plane at the tip of the GSG probes. In this study, transmission behavior was used to quantify the changes occurring in the various structures based on loss (amplitude) and phase changes in the transmission performance. The combination of loss and phase change can provide information on the changes associated with resistance and inductance associated with the metal and capacitance associated with the dielectric in the DUTs before and after thermal stresses are applied to the DUTs.

Figure 4 shows the insertion losses (S_{21}) magnitude, in the 0.7 GHz to 40 GHz frequency range, as a function of the number of thermal cycles for the prototypical TSV samples.[d] Similar to the S_{11} data for the planar samples, the S_{21} losses increased with increasing number of thermal cycles. Notice that the S_{21} magnitude profile of the device under test after thermal cycling is much different from the 'as-received' sample. For instance at 40 GHz, after 2000 thermal cycles the TSV chain's insertion loss degraded by a factor of about 2.4 from its unstressed state (as-received). Furthermore, the degree of deviations of the subsequent traces from the 'as-received trace' shows a strong RF frequency dependence. Specifically, the deviations at frequencies below 5 GHz are much smaller than at higher frequencies, with a very significant roll-off above 25 GHz. Additionally, the RF trace signature is observed to change most drastically after 1000 cycles. The deterioration in the RF characteristics of the TSV interconnect is a clear indication of the damage that have occurred in the structure, such as voids, cracks, delamination and much more, with thermal cycling. The drastic change in the insertion loss profile after 1000 cycles suggests the onset of additional or the exacerbation of existing reliability problems within the DUT.

The frequency dependence of the S_{21} signal deterioration suggests that the broadband RF probes different regions of the DUT, in a manner determined by the 'skin effects'. The penetration depth of the incident electromagnetic waves at the operating frequencies can be calculated using equation 1, where δ is the skin depth, ρ is resistivity of the conductor; ω is angular frequency of current = 2πf and μ is absolute magnetic permeability of the conductor. $\delta = \sqrt{\frac{2\rho}{\omega\mu}}$ [1]As discussed above, we can distinguish between main types of failures, those responsible for signal losses at below 10 GHz and those above 20 GHz.

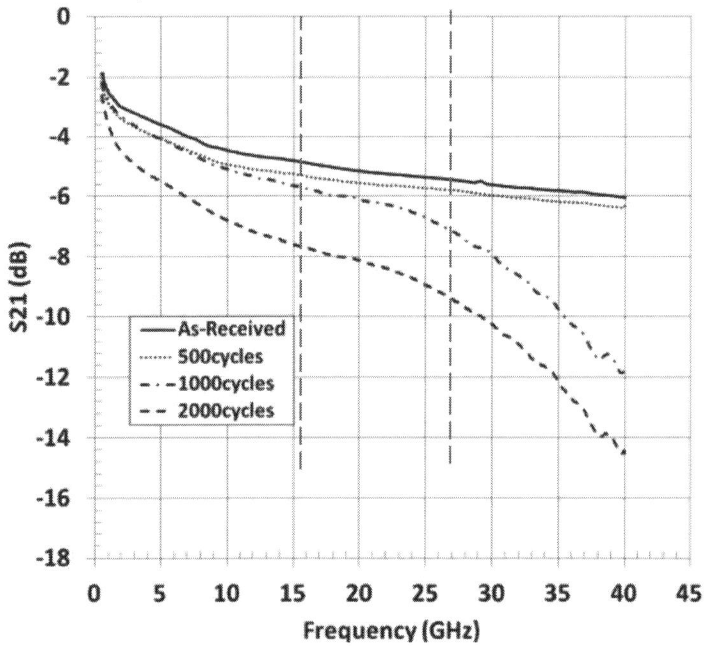

Figure 4. Effect of the number of thermal cycles on the magnitude of the transmission coefficient (S21) as a function of frequency for a single test sample.

For interconnects, such as TSVs, the RF-induced current at high frequencies (greater than 10 GHz), is concentrated at the metal-dielectric interface; i.e., the interface between the Cu and the surrounding barrier (Ta/TaN) and isolation liners (SiO_2) due to the skin depth phenomenon. Thus, a higher frequency electromagnetic wave penetrates less into a TSV, than a low frequency wave.

Figure 5 compares the cross-sectional FIB-SEM images of sections of Cu filled TSV of "as-received" sample to that after 1000 thermal cycles, to show the impact of the thermal cycling on the materials of construction. The losses below 20 GHz suggest that there may be pre-existing small defects in the body of the Cu-fill (as shown in Figure 5a). After 1000 cycles, these defects grow and coalesce, as shown in Figure 5b, and significant S_{21} deviation and roll off. Defect types such as cracks and delamination can occur at the bulk Cu-fill–isolation layer interface, and contribute significantly to the S21 signal loss above 20 GHz.

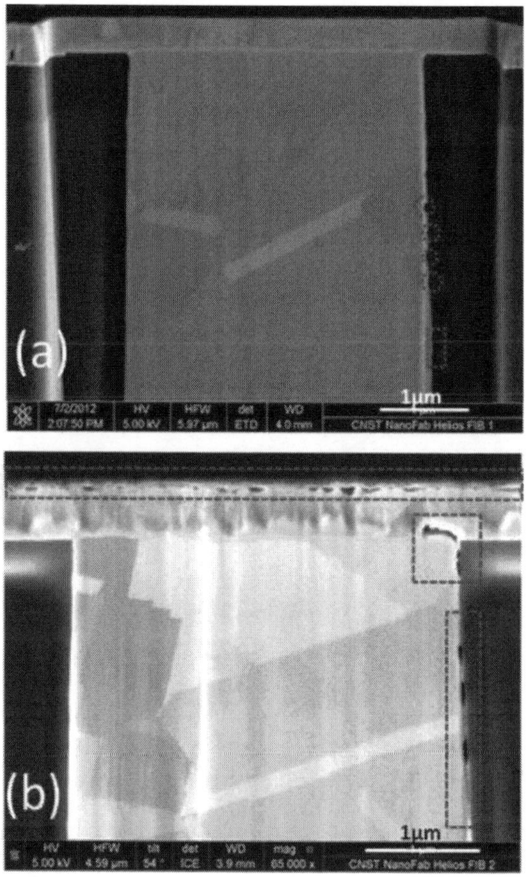

Figure 5. Defects within the Cu filling of the TSV. (a) As-received (b) After 1000 cycles.

Changes in signal propagation characteristics (e.g., S11 amplitude and phase) of device under test are related to the changes in the characteristic impedance of the device.[2] In analogy to a microstrip transmission line, the dielectric constant of the medium surrounding the signal carrying lines determines the phase velocity of the electromagnetic waves through the devices under test.[9] Figure 6 shows the evolution of the signal phase (phase velocities) in the Cu filled TSVs, as a function of the number of thermal cycles endured, in the prototypical 3D-ICs studied. Below 500 cycles, the phase constants are similar and linear up to 40 GHz. Above 1000 cycles, the phase constant is different and starts to show linear and non-linear behaviors. At the high thermal cycles, the phase response is linear up to 15 GHz. The slope remains linear, but changes up to 28 GHz; above 28 GHz, it becomes slightly non-linear. These changes are indicative of significant changes in the propagation velocity of electromagnetic within the devices under test (DUTs) with increasing number of thermal cycles.

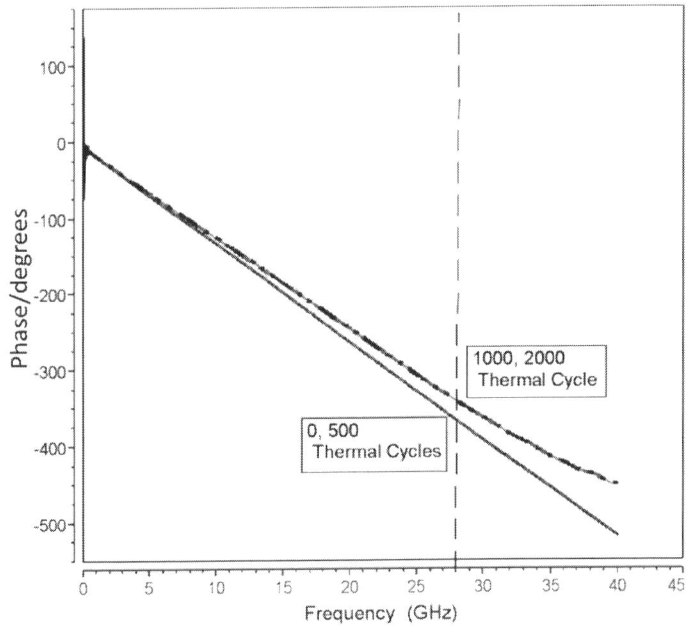

Figure 6. Variation of the Slope of Signal Phase Angle with number of thermal cycles endured by devices under test. The solid trace represents the data for 'as-received' and at 500, while the solid trace represents that data for 1000 and 2000 thermal cycled DUTs respectively. The vertical dashed line at 28 GHz represents the transition from linear to non-liner phase shift for the 1000 and 2000 cycle DUTs.

In Figure 6, the change in velocity suggests an increase in effective dielectric constant of the dielectric surrounding the TSV. We postulate that such increases in dielectric constant could be due to coalescing and migration of small voids in the dielectric. In the extreme, these coalescing voids could manifest as de-bonding between interfaces to introduce air-pockets. Such delamination has indeed been observed in our work.10 The delamination could allow current to leak into the silicon substrate; the non-linear region of the phase data suggest that signal dispersion is in fact present in the high thermal cycle DUTs. Signal dispersion occurs when high resistance and/or inductive or capacitive effects become present, and may be taken as evidence of mechanical failure of the barrier / isolation liner around the Cu-fill.

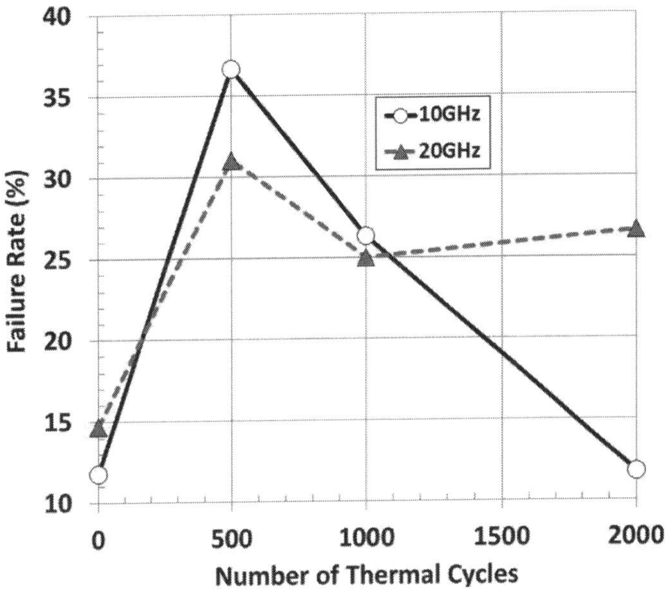

Figure 7. Reliability analysis data showing the relationship between failure rate and the number of thermal cycles at frequencies of 10 and 20 GHz.

Figure 7 compares the cumulative failure rates as a function of number of endured cycles at 10 GHz and 20 GHz respectively. Inspection of the figure shows that the failure rates exhibit different trends at the different RF frequencies. It is instructive that at lower frequencies we look deeper into the Cu-filling the TSV, while at the higher frequencies we can only access the Cu- barrier interface. From equation 1, at 10 GHz we probe up to 656 nm into the Cu-fill from of the Cu-barrier interface, while at 20 GHz we access up to 464 nm from the Cu-barrier interface. Thus, qualitatively, the failure rates at lower frequencies (less than 10 GHz) represent more of what is happening within the bulk of the Cu-fill, while the data at 20 GHz represents predominantly the events closer to the Cu-isolation layer interface, such as cracks and delamination. The insertion loss observed from 500 to 1000 cycles suggests the aggregation of pre-existing defect in bulk Cu fill in response to stress build up. However, at 2000 cycles the insertion loss increases further, which indicates the acceleration of the prevailing active failure mechanisms, especially at the copper-isolation liner interface, and / or the introduction of a different failure mechanism between 1000 and 2000 cycles.

This is further supported by the multiple slopes in the cumulative failure plots for 10 GHz (Figure 8a) and 20 GHz (Figure 8b) respectively. The distributions of the data show at least three different slopes, at both frequencies. This suggests competing failure modes and mechanisms within the TSV in response to the stress buildup from the thermal cycling.11 Identifying and resolving these competing mechanisms is the subject of current ongoing research.

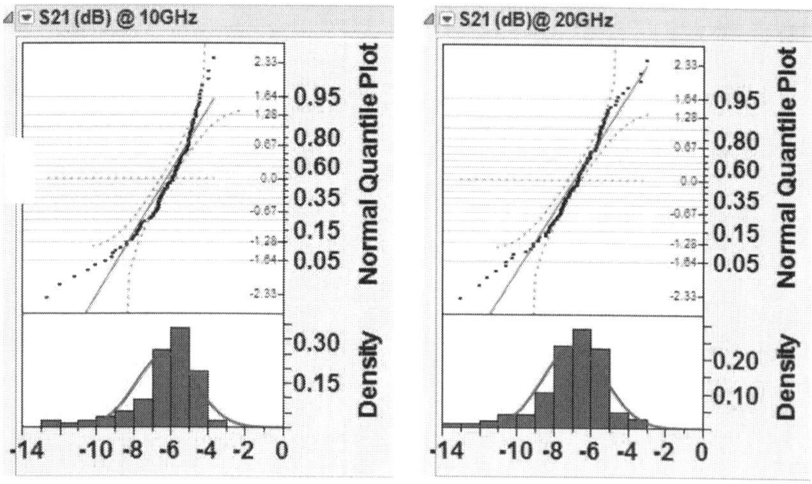

Figure 8. Statistical analysis showing the distribution of all the data points, and are presented as normal quantile and histogram plots at (a) 10 GHz and (b) 20 GHz, respectively.

CONCLUSIONS

In this paper, we have described the failure mode information inherent in high frequency electromagnetic signal loss data. Specifically, we have shown that it is possible to probe material and structural changes in integrated circuits under various external perturbations with broadband RF, as well as to detect buried artifacts. Additionally, the RF-based techniques described here provide several pieces of information, which when taken together provide a more complete picture of the progressive defect growth in the devices under test.

The impact of external perturbation on the reliability of 3D-ICs can be deduced from the resultant changes in the device impedance, through changes in the signal propagation characteristics (e.g., insertion losses and signal velocity). For example, changes in insertion loss above 20 GHz as a function of thermal cycling point to the formation of defects such as cracks and delamination at the bulk Cu-fill–isolation liner interface. Also, changes in the signal velocity suggest the coalescing and migration of small voids in the dielectric away from the copper fill, toward the silicon substrate. In the extreme, this migration could result in interfacial de-bonding and allow current leakage into the silicon substrate.

Finally, we have demonstrated that the S11 data from SMM technology can detect in-service material changes, such as degradation and corrosion of metal lines.

FOOTNOTES

- [*]Electrochemical Society Active Member.
- [z]E-mail: yaw.obeng@nist.gov
- *This paper is part of the JSS Focus Issue on Advanced Interconnects: Materials, Processing, and Reliability.*
- [c]Certain commercial equipment, instruments, or materials are identified in this report in order to specify the experimental procedure adequately. Such identification is not intended to imply recommendation or endorsement by the National Institute of Standards and Technology, nor is it intended to imply that the materials or equipment identified are necessarily the best available for the purpose.
- [d]Figures 4 and 5 were reproduced from Reference 1 with authors' permission.
- Manuscript submitted August 28, 2014.
- Revised manuscript received October 27, 2014.
- Published November 7, 2014.

REFERENCES

1. C. Okoro, P. Kabos, J. Obrzut, K. Hummler, and Y. S. Obeng, IEEE Transactions on Electron Devices, 60(6), 2015 (2013).
2. S. Foley, L. Floyd, and A. Mathewson, presented at the Reliability Physics Symposium Proceedings, 1999. 37th Annual. 1999 IEEE International, 1999.
3. K. Daeil, M. H. Azarian, and M. Pecht, IEEE Transactions on Device and Materials Reliability, 9(2), 296 (2009).
4. S. Anlage, V. Talanov, and A. Schwartz, in Scanning Probe Microscopy, edited by Sergei Kalinin and Alexei Gruverman (Springer New York, 2007), pp. 215.
5. H. P. Huber, M. Moertelmaier, T. M. Wallis, C. J. Chiang, M. Hochleitner, A. Imtiaz, Y. J. Oh, K. Schilcher, M. Dieudonne, J. Smoliner, P. Hinterdorfer, S. J. Rosner, H. Tanbakuchi, P. Kabos, and F. Kienberger, Review of Scientific Instruments, 81(11) (2010).
6. C. Okoro, L. E. Levine, R. Xu, K. Hummler, and Y. Obeng, Journal of Applied Physics, 115(24), 243509 (2014).
7. C. Okoro, Y. Yang, B. Vandevelde, B. Swinnen, D. Vandepitte, B. Verlinden, and I. De Wolf, presented at the Interconnect Technology Conference, 2008. IITC 2008. International, 2008.
8. M. van Soestbergen, A. Mavinkurve, R. T. H. Rongen, K. M. B. Jansen, L. J. Ernst, and G. Q. Zhang, Electrochimica Acta 55(19), 5459 (2010).
9. T. C. Edwards, Foundations for microstrip circuit design. (J. Wiley, 1981).

10. C. Okoro, J. W. Lau, F. Golshany, K. Hummler, and Y. S. Obeng, IEEE Transactions on Electron Devices, 61(1), 15 (2014).

11. W. Huang and R. G. Askin, Quality and Reliability Engineering International 19(3), 241 (2003).

2

NOVEL PRINCIPLE OF CONTACTLESS GAUGE BLOCK CALIBRATION

Zdeněk Buchta *, Šimon Řeřucha, Břetislav Mikel, Martin Čížek, Josef Lazar and Ondřej Číp

Institute of Scientific Instruments, Academy of Sciences of the Czech Republic, Královopolská 147, 612 64 Brno, Czech Republic

ABSTRACT

In this paper, a novel principle of contactless gauge block calibration is presented. The principle of contactless gauge block calibration combines low-coherence interferometry and laser interferometry. An experimental setup combines Dowell interferometer and Michelson interferometer to ensure a gauge block length determination with direct traceability to the primary length standard. By monitoring both gauge block sides with a digital camera gauge block 3D surface measurements are possible too. The principle presented is protected by the Czech national patent No. 302948.

Keywords: low-coherence interferometry; gauge block; nanometrology

1. INTRODUCTION

The use of length measuring instruments is a usual part of everyday life and these instruments are used in almost all branches of human activity. In general, a suitable length measuring instrument could be arbitrarily chosen with respect to the measurement range and accuracy needed. The accuracy of a length measuring instrument is ensured and verified by an instrument calibration process.

As for calibration of the length measuring instruments, there is no liberty in choice of a calibration instrument. Due to the basic principle of length metrology, the calibration instrument has to be more accurate, in ideal case by one order more than the instrument which is being calibrated.

In mechanical engineering practice, gauge blocks are at the end of the calibration chain. Gauge blocks are used as mechanical length etalons for an accurate calibration of length measuring instruments such as micrometers and calipers. Gauge blocks are made of steel, metal carbide or ceramics. In all cases, the gauge blocks eventually wear away for many reasons, including the mechanical contact between the gauge block and the calibrated instrument, material and form changes *etc.* For this reason, a regular gauge block calibration is necessary and it is a hot topic in fundamental and industrial metrology [1]. The methodology of gauge block calibration is defined by the international standard EN ISO 3650 describing two principal ways of calibrating gauge blocks.

The first method employs laser interferometry, white-light interferometry or multiwavelength interferometry [2,3]. The gauge block is adhered to the reference surface. Then the length of the gauge block is measured by one of the interferometric techniques mentioned above.

The second one is the contact method. In this case, the gauge block length is compared against a reference gauge block. The measured and reference gauge blocks are probed from both sides with a couple of incremental length gauges.

From a practical point of view, a contactless calibration technique could be a useful alternative to the techniques used nowadays. The main advantage is a possible automation of the calibration process.

The presented optical setup for contactless gauge block measurement is based on double-ended interferometry combining a laser and a white-light source and it is presented as an alternative to similar configurations published before [4–8]. The key advantage of the presented optical setup is its contactless measurement of the absolute gauge block length done as a single-step measurement giving complete information of the gauge block length, without any additional gauge block manipulation or any length comparison with a reference dimension. The measurement is fully automatic, with no operator influence.

Low-coherence interferometry is a powerful diagnostic technique which has become an attractive and a useful method for fast and accurate 3D inspection of macroscopic objects. Typical fields of application of white-light interferometry are contactless surface measurement of macroscopic objects, surface roughness and quality assurance, measurement of the thickness of thin films and optical tomography.

The principle of white-light interferometry stems from classical laser interferometry. The main difference between these two techniques is the type of the radiation source used. In case of white-light interferometry, a low-coherent light emitter (halogen lamp, xenon lamp, superluminescent LED diode) is used instead of a highly-coherent laser.

The principle of white-light interferometry is given by the white-light source. Its coherence length is a few micrometers (typically from 3 μm to 5 μm) [9,10]. The white-light source coherence length defines the range of the reference mirror movement in which the reference beam and the measuring beam interfere. This property predetermines the white-light interferometer to be an indicator of the reference interferometer arm and the measuring interferometer arm balance.

2. EXPERIMENTAL SETUP

The presented optical setup combines a Michelson interferometer and a Dowell interferometer [11], placed in the reference arm of the Michelson interferometer. The principle of the measurement is illustrated inFigure 1. A parallel beam, generated by a white-light source, is divided into two parts by semireflecting mirror 1. The resulting measuring beam goes through a couple of compensating plates CP1 and CP2 and after that is reflected back by a reference surface RS.

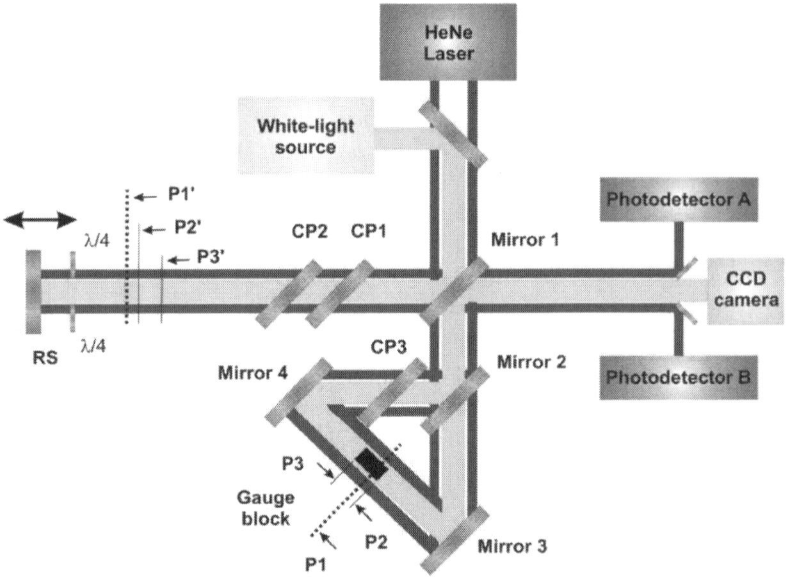

Figure 1. Optical setup for gauge blocks measurement. CP1, CP2 and CP3 are compensating plates, RS is a reference surface, λ/4 is a retardation quarterwave plate.

The resulting reference beam of the Michelson interferometer plays the role of the primary beam for the Dowell interferometer. This beam is divided by mirror 2 into two beams going in the Dowell interferometer in opposite directions. One of them passes through compensating plate CP3 and is reflected by mirror 4 onto a gauge block face. The other beam passes through the beamsplitter (mirror 2) and then is reflected by mirror 3 onto the other face of the gauge block. One part of both mentioned beams illuminating the gauge block is reflected back by the gauge block faces and the other part of both beams pass alongside the gauge blocks.

At the output of the Michelson interferometer, there are five beams which could possibly interfere—the first one is the measuring beam of the Michelson interferometer reflected by the reference surface RS, then there is a pair of beams reflected by the gauge blocks and a pair of beams that pass alongside the gauge blocks.

The principle of the measurement is based on low-coherence interferometry taking advantage of low-coherence properties of the broadband light source. In the range of the movable reference surface shift, there are three positions where the white-light beams interfere.

In the block diagram of the experimental setup shown in Figure 1, these positions are marked as P1′, P2′ and P3′.

In the P1′ position, the Michelson interferometer measuring beam interferes with the pair of beams passing alongside the measured gauge block. In fact, this is equivalent to a configuration with a mirror in a position marked as P1 (see Figure 1). P1 is at the mean optical path length of the ring interferometer, *i.e.*, the equidistance plane for both clockwise and anti-clockwise propagation after mirror 2, taking into account CP3. For the gauge block length measurement, it plays the role of the reference position.

As for positions P2′ and P3′, the Michelson interferometer measuring beam interferes with the beams reflected by the gauge block faces (marked as P2 and P3 in Figure 1). Then, the measured gauge block length is equal to the sum of distances between the measuring positions P2′ and P3′ of the reference position P1′.

Figure 2 shows a sketch of a screenshot recorded by the CCD camera. There are two interference areas—the beams in the area no. 1 interfere in reference surface positions P2′ and P3′ and in the area no. 2, the beams interfere in reference surface position P1′. During a gauge block measurement procedure, only the interference signal from the relevant area of interference is recorded. The recorded data are used for complex gauge block analysis, namely gauge block length measurement, gauge block face surface measurement and for interference fringe shape analysis to evaluate the gauge block flatness distortion.

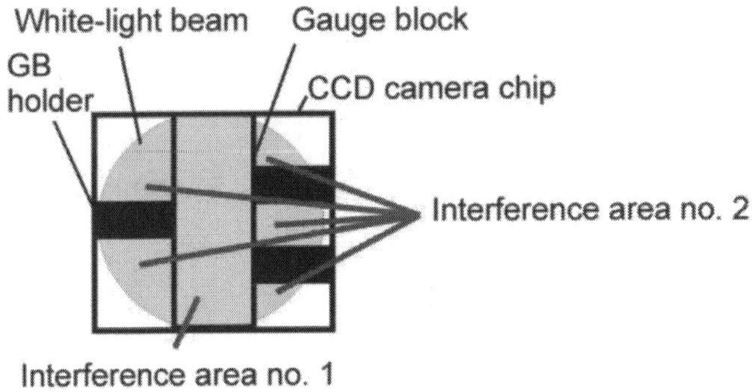

Figure 2. Sketch of the screenshot recorded by the CCD camera.

For illustration, an interference detecting signal (IDS) record is shown in Figure 3. The signal was recorded during a pilot gauge block measurement experiment and it represents a correlation coefficient value resulting from a comparison of two sequential samples of the output beam intensity. In the designed system, this signal is used for an automatic detection of the measuring positions. In the case shown in Figure 3, a gauge block with the length 17 mm was measured.

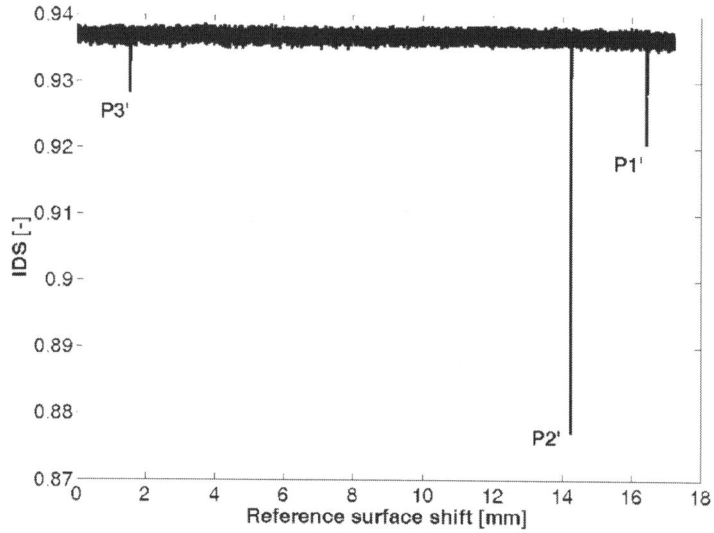

Figure 3. The Interference Detecting Signal (IDS) recorded during a pilot experiment with gauge block measurement. In the case shown in this figure, a gauge block with the length 17 mm was measured.

The gauge block measurement process is controlled by special software which was designed for the positioning control of the reference surface RS and the online acquisition of the signals from the photodetectors and a CCD camera. The measurement control process is based on monitoring the IDS signal during a fast positioning of the reference surface. The interference of the output beams is detected as a drop of the IDS signal. In this case, the motorized stage is switched off and the interference signal is measured by means of a PZT driven stage. The measuring positions P1′, P2′ and P3′ are localized by a central point detection algorithm implemented in the measurement control software. The measured values of positions P1′, P2′ and P3′ has to be corrected to respect the phase shifts caused by light beam reflections and parameters of the optical components used. The reference surface RS was made from the same material and by the same way like standard gauge block. Due to this fact, the estimated phase step on reflection is for positions P2′ and P3′ the same for the reference surface and for the gauge block and for steel gauge blocks it doesn't need any other correction. In the case of position P1′, there is no additional reflection on the gauge block surface. This makes the measured P1′ position different from the right one and the total gauge block length seems to be longer than it in fact is. For P1′ correction, the missing phase step on reflection has to be taken into account—the phase step on reflection is 160° for the steel gauge block, which is the mean value calculated for the visible part of the optical spectrum using the optical constants published in [12]. Furthermore, the measured gauge block length (GBL) is calculated by the Equation (1):

$$GBL = |P1' - P2'| + |P1' - P3'| \tag{1}$$

By using of the CCD camera as a photodetector, this measuring principle allows measurement of the gauge block length in a number of points for verification of gauge block flatness and parallelism.

For precise gauge block measurements, it is necessary to acquire the measuring positions with nanometer range precision. In the designed experimental setup, this is ensured by an incremental laser interferometer which complements the low-coherence interferometer described above. As a source of laser radiation, a 633 nm He-Ne laser is used. Its output beam is collimated into an optical fiber, then split by a 50/50 fiber splitter and the resulting two laser parallel beams are used for the incremental interferometric measurement as shown in Figure 1. In general, it works as the classic Michelson interferometer. In the reference arm of the interferometer, the laser beams pass alongside the measured gauge block and at the output of the interferometer, they are combined with the laser beams passing through the measuring interferometer arm. The detection unit (DU) used as a photodetector A and B (see Figure 1) receives two associated beams from the measuring and reference arm of the interferometer [13].

The measurement setup is shown in Figure 4. It is arranged on a massive base plate made of gray iron with a thermal expansion coefficient of 12.3×10^{-6} K^{-1}. The optical components are put into 3″ gimbal transmitting mounts TRANS 90G (OWIS) optimized for a maximum aperture. These transmitting mounts are combined with goniometers with a range of adjustment of ±10° designed and made for precise rotation of the optical components around a defined optical axis. The final adjustment can be fixed with a screw. The used transmitting mounts are made from aluminum alloy with a declared thermal expansion coefficient of $23.5 \times 10^{-6}\ K^{-1}$ and the gauge block holder and the reference surface holder are made from steel with a thermal expansion coefficient of $10.6 \times 10^{-6}\ K^{-1}$.

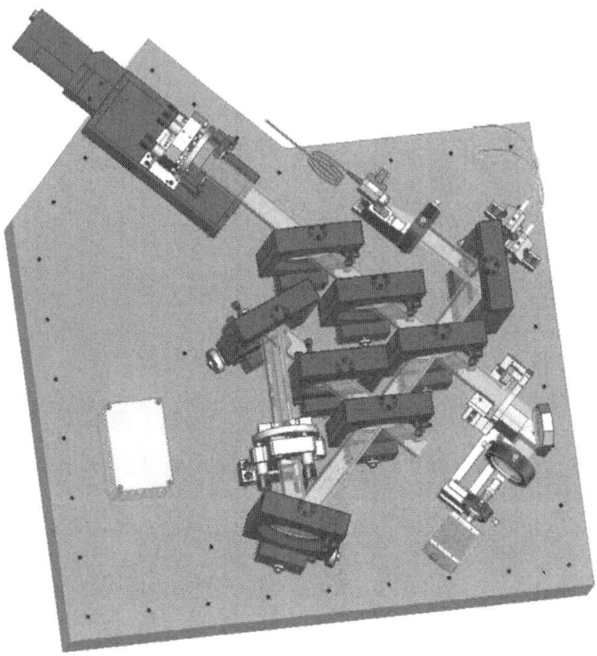

Figure 4. Sketch of the practical realization of the designed experimental setup.

For the measurement process, the gauge block is placed by an automatic handling system into a gauge block holder supplemented by three PZT screw actuators (Thorlabs) used for gauge block adjustment before its measurement. The automatic handling system is designed for a set of 126 gauge blocks (0.5 mm to 100 mm) to allow the automatic contactless calibration of the complex gauge block set without a human operator.

The positioning of the reference surface is ensured by a combination of a LIMES 80-100-HiDS (OWIS) motorized positioning stage and a reference surface holder supplemented—like the gauge block holder—by three PZT screw actuators (Thorlabs). The travel of the

Limes stage is 105 mm with a spindle pitch of 1 mm and 2,000 steps per revolution. The screw actuator with PZT provides 4 mm of manual coarse travel via a 0.25 mm pitch leadscrew. The 150 V internal piezo stack provides 15 μm of open-loop piezo travel and an 18-bit DA converter used in control electronics allows extremely fine positioning of the reference surface.

The combination of these two types of actuator meets the requirement for reference surface positioning in the 100 mm range with subnanometer resolution. In addition, using three PZT screws allows an online correction of the main stage pitch and yaw deviations.

Except for the main part of the optical setup, the measurement system was designed as a modular device. The sources of white-light and laser radiation and the detection system are made as separate modules to allow precise adjustment out of the system.

The base plate with the optical setup, shown in Figure 4, is oriented to the vertical position to respect the international standard EN ISO 3650 which requires the vertical calibrating position for gauge blocks of up to 100 mm in length.

The whole optical setup is covered by a double shielded housing for minimization of refractive index of air fluctuations. The value of the refractive index of air is calculated from measured atmospheric parameters by means of the Edlen formula and it is used for an online correction of the HeNe laser wavelength with 1×10^{-7} precision [14,15]. In addition to an ambient temperature measurement, there are three temperature sensors monitoring the base plate temperature for corrections of the optical path dilatation during the gauge block measurement process.

The described method was designed to be an alternative for the methods presently used for gauge block calibration. In comparison with the current methods, the key advantage of the presented optical setup is its contactless measurement of the absolute gauge block length, which is fully automatic, with no operator influence. Because of the automatic measurement process, a set of gauge blocks could be measured one by one without delay caused by waiting for temperature stabilization of each block. As for the time required for the diagnosis of one gauge block, it depends mainly on the set motorized stage and PZT scanning speed and this has to be set with respect to the electronics used (CCD cam frame rate, laser interferometer detection unit sensitivity and maximal sampling rate,...) and detection techniques (in our case, coherent correlation technique [16] for white-light fringe center detection and homodyne detection technique for laser interferometers). In the experimental setup in our lab, the diagnostics of a gauge block of 20 mm length takes about 120 s. During the pilot experiments, the repeatability of the measurement was tested. Generally, it depends on the measurement conditions, setup mechanical stability and also on used central fringe detection technique. For the described experimental setup, a standard deviation of six consecutive P1′ position measurements was 7.3 nm.

From the metrological point of view, the method's relative uncertainty is a crucial factor. The most important factors making up the method's relative uncertainty are the thermal expansion coefficients of the materials used (gauge block $10.6 \times 10^{-6}\,K^{-1}$, base plate $12.3 \times 10^{-6}\,K^{-1}$, gauge block holder and reference surface holder $10.6 \times 10^{-6}\,K^{-1}$ and transmitting mounts $23.5 \times 10^{-6}\,K^{-1}$) and the method of measurement of the refractive index of air.

The uncertainty of the refractive index of air is calculated as the combined standard uncertainty by the values presented in Table 1. The $U(x_i)$ is a standard uncertainty of an atmospheric parameter measurement, c_i is a relative change of an refractive index of air caused by change of an atmospheric value and $U_i(n)$ is a standard uncertainty of an atmospheric parameter.

Table 1. Refractive index of air—standard uncertainty.

	x_i	$U(x_i)$	$c_i = \partial l/\partial x_i$	$U_i(l)$ / rel.
Atmospheric pressure	99 kPa	10	3E-9 /Pa	2.7E-08
Air temperature	20°C	0,01	1E-6 /K	1.0E-08
Relative humidity	60%	1	1E-8 /%	1.0E-08
CO_2 concentration	700 ppm	10	1.5E-10/ppm	1.5E-09
Edlen formula	1.0002650	2.E-08	1	2.0E-08
RIA relative uncertainty				3.6E-08

The measurement uncertainty given by mechanical parameters of the experimental setup is calculated as a combined standard uncertainty using the values presented in Table 2.

Table 2. Uncertainty part given by mechanical parameters of the experimental setup

	x_i	$U(x_i)$	$c_i = \partial l / \partial x_i$	$U_i(l)$/ rel.
Gauge block thermal expansion	20°C	0.01	10.6E-6	1.1E-07
Air temperature	20°C	0,01	12.3E-6	1.2E-07
Mechanical components thermal expansion (gauge block holder, reference surface holder)	20°C	0,01	10.6E-6	1.1E-07
Total mechanical relative uncertainty				1.9E-07

In our setup version, the estimated relative uncertainty value is 2.0×10^{-7}.

3. CONCLUSIONS

This paper presents a novel principle for contactless gauge block measurement using a combination of low-coherence interferometry and laser interferometry. The experimental setup combines a Dowell interferometer and a Michelson interferometer to ensure a gauge block length determination with direct traceability to the primary length standard. This setup was designed for contactless complex gauge block analysis providing information about gauge block length, gauge block faces surface profile (e.g., indication of scratches) and by analysis of the the interference fringes shape, also about the gauge block edge flatness distortion. The designed setup is supplemented by an automatic handling system designed for a set of 126 gauge blocks (0.5 mm to 100 mm) to allow the automatic contactless calibration of the complex gauge block set without a human operator. The described method was recently protected under Czech national patent No. 302948.

ACKNOWLEDGMENTS

This work was supported mainly by Grant Agency of CR project GP102/09/P293 and partially by GAP102/11/P819 and GP102/10/1813. The pilot set-up of the method was funded by Ministry of Industry and Commerce, project No: 2A-1TP1/127. The background of the research was supported by the European Commission and Ministry of Education, Youth, and Sports of the Czech Republic project No. CZ.1.05/2.1.00/01.0017 and projects LC06007.

REFERENCES

1. Doiron, T.; Beers, J. *The Gauge Block Handbook*, NIST Monograph 180 with Corrections; 145. June 1995, pp. 1–143. Available online: http://emtoolbox.nist.gov/Publications/NISTMonograph180.asp/ (accessed on 2 March 2012).

2. Decker, J.E.; Miles, J.R.; Madej, A.A.; Siemsen, R.F.; Siemsen, K.J.; de Bonth, S.; Bustraan, K.; Temple, S.; Pekelsky, J.R. Increasing the range of unambiguity in step-height measurement with multiple-wavelength interferometry—application to absolute long gauge block measurement. *Appl. Opt* **2003**,*42*, 5670–5678.

3. Bönsch, G. Automatic gauge block measurement by phase stepping interferometry with three laser wavelength. *Proc. SPIE* **2001**, *4401*, 1–10.

4. Khavinson, V.M. Ring interferometer for two-sided measurement of the absolute lengths of end standards. *Appl. Opt* **1999**, *38*, 126–136.

5. Ishii, Y.; Seino, S. New method for interferometric measurement of gauge blocks without wringing onto a platen. *Metrologia* **1998**, *35*, 67–73.

6. Abdelaty, A.; Walkov, A.; Abou-Zeid, A.; Schödel, R. PTB'S prototype of a double ended interferometer for measuring the length of gauge blocks. Proceedings of the Simposio de Metrologia, Queretaro, Mexico, 27–29 October 2010; pp. 1–6.

7. Dobosz, M.; Iwasinska-Kowalska, O. A new method of non-contact gauge block calibration using a fringe-counting technique: Theory basis. *Opt. Laser Tech* **2010**, *42*, 141–148.

8. Iwasinska-Kowalska, O.; Dobosz, M. A new method of non-contact gauge block calibration using a fringe-counting technique: Experimental verification. *Opt. Laser Tech* **2010**, *42*, 149–155.

9. Buchta, Z.; Mikel, B.; Lazar, J.; Číp, O. White-light fringe detection based on novel light source and color CCD camera. *Meas. Sci. Tech* **2011**, *22*, 094031.

10. Ikonen, E.; Kauppinen, J.; Korkolainen, T.; Luukkainen, J.; Riski, K. Interferometric calibration of gauge blocks by using one stabilized laser and a white-light source. *Appl. Opt* **1991**, *30*, 4477–4478.

11. Dowell, J.H. Improvements in or relating to Interferometers for Determination of Length, British patent No. 555672. 1943.

12. Karlsson, B.; Ribbing, G.C. Optical constants and spectral selectivity of stainless steel and its oxides. *J. Appl. Opt* **1982**, *53*, 6340–6346.

13. Číp, O.; Petrů, F. A scale-linearization method for precise laser interferometry. *Meas. Sci. Technol* **2000**,*11*, 133–141.

14. Číp, O.; Buchta, Z.; Petrů, F.; Lazar, J. On-line monitoring of the refraction index of air for ultra-precise length measurement in the nano-world. Proceedings of 8th IEEE Africon Conference, Windhoek, Namibia, 26–28 September 2007; pp. 294–298.

15. Lazar, J.; Číp, O.; Čížek, M.; Hrabina, J.; Buchta, Z. Suppression of air refractive index variation in high-resolution interferometry. *Sensors* **2011**, *11*, 7644–7655.

16. Ailing, T.; Chunhui, W.; Zhuangde, J.; Hongjun, W.; Bingcai, L. Study on key algorithm for scanning white-light interferometry. *Proc. SPIE* **2008**, *71552N*, 1–7.

3

CARBON NANOTUBE AFM PROBE TECHNOLOGY

Z. W. Xu[1], F. Z. [a] - Fang[1] and S. Dong[2]

[1] State Key Laboratory of Precision Measuring Technology & Instruments,Centre of MicroNano Manufacturing Technology, Tianjin University, China

[2] Center for Precision Engineering, Harbin Institute of Technology, China

1. INTRODUCTION

The invention of atomic force microscopy (AFM) is having a great impact on various areas, such as nano metrology, materials science, surface science and biology (Binnig et al., 1986). The lateral resolution of AFM is mainly determined by the probe's shape and physical property, especially the geometry and dimension of the probe end. Conventional AFM probe is pyramidal shape by micro-fabrication. The pyramidal probe would result in image resolution degradation by severe probe broaden effect, especially for the structures with higher aspect ratio, such as gratings and structures in MEMS.

To broaden the AFM applications, researchers pursue new kind probes that have longer lifetime, higher resolution, and better mechanical property. Carbon nanotubes (CNT) show many excellent properties, such as, high aspect ratio, high Young's modulus, excellent elastic buckling property, and electrical and thermal conductivity (Iijima,

1991). The above characteristics make carbon nanotube be ideal as probes in AFM. Carbon nanotubes have demonstrated considerable potential as AFM probes after the first CNT AFM probe was invented in 1996 (Dai et al., 1996).

This chapter would introduce the history of carbon nanotube AFM probes, including the CNT probes' farbication and configuration optimization, the CNT probes' image artefact and its elimination study, the applications of these new kind probes and researches to improve their performance.

2. CNT PROBES' FABRICATION

There are two key processes in the fabrication of carbon nanotube AFM probe: attachment and modification. Firstly, the CNT must be fixed to the end of ordinary AFM probe to be the new imaging point; and then configuration or function modification to optimize the CNT probe's properties.

2.1. CNT Attachment

2.1.1. Direct Manipulation Method

Since carbon nanotubes were first applied as AFM probes in 1996 by Dai (Dai et al., 1996), various methods have been developed for their fabrications. The earliest method was to

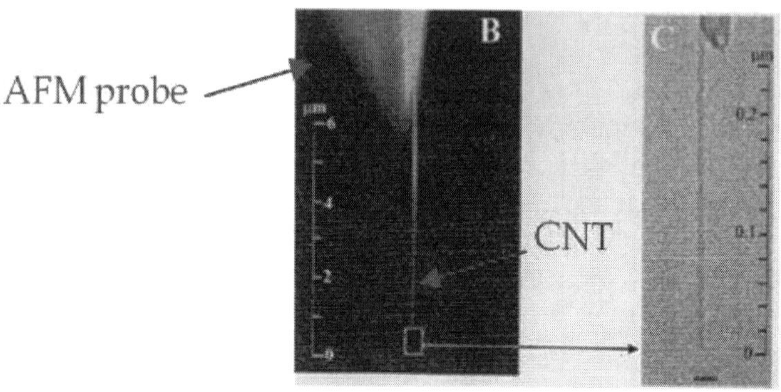

Figure 1. SEM and TEM images of the first CNT AFM probe (Dai et al., 1996)

employ precise manipulation by picking and sticking a multi-walled carbon nanotube (MWNT) bundle to the silicon probe with an acrylic adhesive under direct view of optical microscope. This method can not be employed to well control the CNT probe's orientation. Further research extended this method using nanomanipulator in scanning electron microscope (SEM), and nanotube with smaller diameter can be selected and fixed to Si probe in SEM by electron beam induced carbon deposition (Nishijima et al., 1999) or Pt depostion (Fang et al., 2009).

Welding method was then developed for fixing CNT to Si probe end (Fang et al., 2007 ; Stevens et al., 2000). First, silicon probe and carbon nanotube were brought into a close distance by manipulating two microtranslators under direct view of inverted optical microscope. When carbon nanotube and silicon probe were in close proximity, an electric field of less than 20V was applied between them. The nanotube was attracted to the silicon probe and favorably aligned with the apex of silicon tip. Then the microtranslators are manipulated to make the protruding nanotube contact with the apex of silicon probe. The applied voltage is further increased between them to 30–60V until the nanotube was energetically disassociated by applying voltages and weld to the end of silicon probe. The carbon nanotubes grown in CVD method have small defects, at which points the resistance locally heats the nanotube until it is oxidized and divided. Fig. 2 is the schematic illustration of fabrication method.

The main drawback of the direct manipulation method is time consuming, and it is not an appropriate method for wafer-scale fabrication.

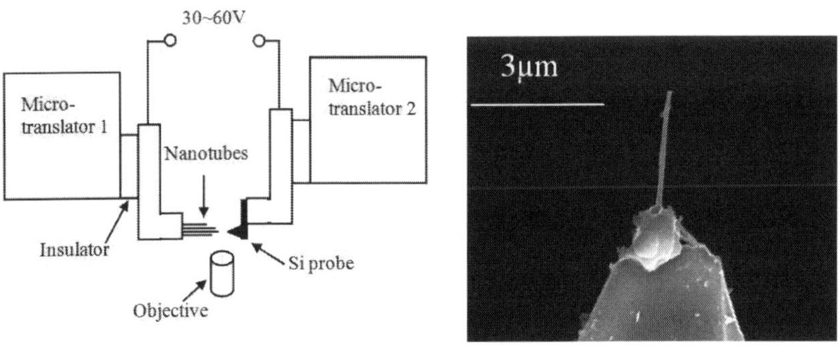

Figure 2. Schematic illustration of welding process and CNT probe fabricated (Fang et al., 2007)

2.1.2. Chemical Vapor Deposition (Cvd) Method

Chemical vapor deposition (CVD) method was used to grow carbon nanotubes on a catalyst deposited silicon probe surface in 1999 (Hafner et al., 1999). The CNT's growth length can be controlled by adjusting the growth time (Edgeworth et al., 2010). Ye reported an innovative approach that combined chemical vapor deposition with nanopatterning and traditional silicon micromachining technologies to large-scale grow CNT probe through effective nanocatalyst protection and release before and after the microfabrication (Ye et al., 2004).

The key advantage of the CVD direct growth method is the possible route to batch fabrication. It was predicted that nanotube bundle AFM probes could achieve reproducibly be wafer-scale fabrication by this technique in the near future [Wilson & Macpherson, 2009]. While, the reproducible production of CNT probes with individual single-walled carbon naontube (SWNT) is still the great challenge. How to accurately control the CNT probe's orientation becomes a key problem for the CVD method.

2.1.3. Pick-Up Method

In order to fabricate SWNT AFM probe, pick-up method was proposed in 2001 (Hafner et al., 2001). The SWNT could be picked up by the ordinary AFM probe with van der Waals forces during AFM probe scanning the SWNT substrate with nanotubes grow upwards. The pick-up method is flexible, while this method needed isolated and vertically aligned carbon nanotube samples.

2.1.4. Other Approaches

Approaches have also been proposed to fabricate CNT probe in CNT solution under external fields, such as, using dielectrophoresis (Tang et al., 2005), alternating magnetic field (Hall et al., 2003). Tamara directly grew CNTs on the apex of AFM probe by selective heating of the catalyst under the microwave irradiation in the presence of ethanol vapor (Tamara et al., 2010). CNT probes were assembled by transplanting a CNT bearing polymeric carrier to a AFM cantilever (Kim et al., 2009).

2.2. Configuration Modification

There are three key factors that determine the resolution and properties of the CNT probe, i.e., radius r, length L, and orientation θ (CNT orietation angle with respect to the sample surface). It is vital that a precise control of these parameters allows the optimization of CNT probe to meet different applications' demand.

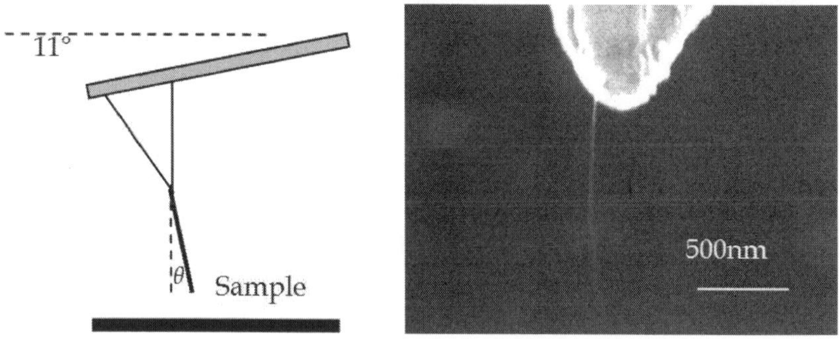

Figure 3. Illustration of the CNT probe's thermally induced oscillations

A nanotube probe would be very gentle in lateral direction if the nanotube's aspect ratio (*L*/*r*) is too large. Fig. 3 shows a scanning electron microscope (SEM) image of a long CNT AFM probe. The blurring of the lower portion of the CNT is due to the thermally induced oscillations.

The lateral force constant (*kl*) of the CNT probe can be derived as the following

$$k_l = \frac{3YI}{L^3} = \frac{3\pi Y}{4} \cdot \frac{r^4}{L^3} \tag{1}$$

where *I* is the stress moment over the cross-section of the nanotube, *Y* is the Young's modulus of the carbon nanotube. The CNT probe bending response will dominate for a probe angle of even a few degrees with respect to the sample surface.

2.2.1. Orientation Modification

The CNT probe's orientation should be vertical to sample surface theoretically for high resolution imaging. However, it is hard to accurately control the CNT probe's orientation during the CNT attachment process for most methods.

External electric field method was effective for CNT probe's orientation control (Fang et al., 2007). In the welding method, nanotube was attracted to the opposite Si probe and favorably aligned with the apex of probe under external applying voltage, as shown in Fig. 4. The attraction is due to the induced dipole moment in the nanotubes under external voltage. The largest strength of the electric field appears at the probe's apex, which will cause the CNT alignment with the apex. In the plasma enhanced chemical vapor deposition (PECVD) method, the CNT probe's orientation was controlled by an electric field present in the plasma discharge to align the nanotubes parallel to the electric field (Ye et al., 2004).

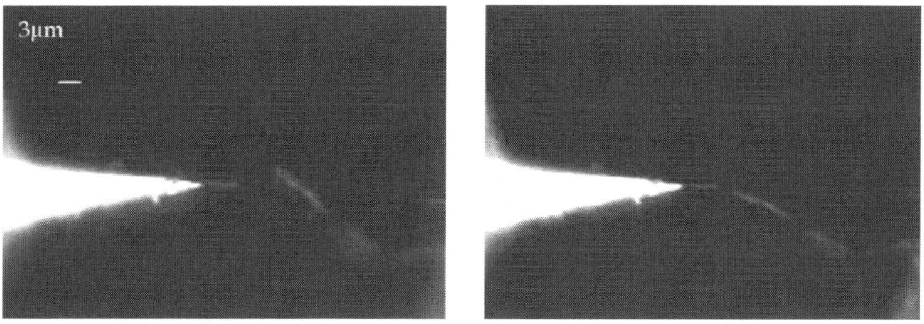

a) Without the voltage   b) With a voltage of 30V

Figure 4. The protruding nanotube favorably aligned with the probe apex (Fang et al., 2007).

Focused ion beam (FIB) irradiation method was recently proposed to modify the CNT probe orientation with high accuracy and reproducibility (Deng et al., 2006; Park et al., 2006). For the CNT probe's orientation alignment process, repeated single scans of the ion beam imaging were applied. The CNT probe was gradually aligned towards the FIB irradiation direction under every FIB scan and finally aligned parallel to the FIB irradiation direction, as shown in Fig. 5. The variables that affect the FIB-CNT interaction process include ion accelerating voltage, ion beam current, exposure time, and ion scan size (or the scan pixel density). Optimized combinations of these variable values can be made to achieve acceptable results (Fang et al., 2009).

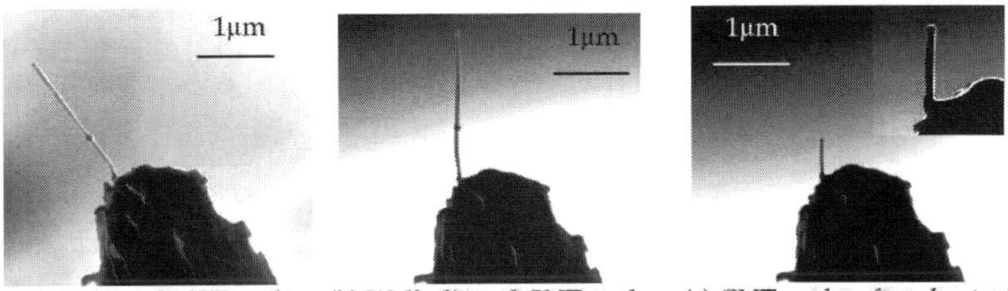

(a) Misaligned CNT probe (b) Well-aligned CNT probe (c) CNT probe after shortening

Figure 5. Nanotube probe's orientation and length are optimized by FIB irradiation and milling processes. The FIB alignment parameters in Figure 5(b) are 30kV and 50pA in the ion acceleration voltage and the ion current, respectively (Fang et al., 2009).

The CNT orientation changes are mainly due to the strain induced by the FIB irradiation and the CNT's excellent plastic ability. Experimental and simulation results have demonstrated that the carbon nanotubes can be considered as self-healing materials under electron or ion irradiation (Krasheninnikov & Nordlund, 2010). Two mechanisms govern the CNT defect annealing (Krasheninnikov et al., 2002). The first mechanism is vacancy healing through dangling bond saturation and by forming nonhexagonal rings. The second one is the migration of carbon interstitials and vacancies, followed by Frenkel pair recombination.

2.2.2. Length Modification

The CNT probe's length should be chosen and modified for different applications. Shortening the CNT probe can decrease the thermal oscillation amplitude. Therefore, short CNT probe is required for the CNT probe's high resolution imaging or applicatons with high lateral force constant requirement, such as friction study (Lai et al., 2010). On the contrary, long CNT probe is useful for the high aspect ratio imaging, e.g., biological samples.

The first method to shorten the CNT probe was realized on AFM by electrical etching (Dai et al., 1996). Then the method of using electron bombardment under the electric field was applied in the direct manipulation methods under optical microscope or SEM, which can observe and control the shortening process in real-time (Fang et al., 2007).

A new 'nanoknife' method was proposed to precisely cut and sharpen CNT probe by local vaporization of carbon resulting from Joule heating (Wei et al., 2007). The 'nanoknife' was a short carbon nanotube adhered to a metal tip. The 'nanoknife' cutting process is controllable and repeatedly, as shown in Fig. 6. In the cutting process, a DC voltage of 5–10 V was applied between the CNT probe and the 'nanoknife', the 'nanoknife' was then manipulated to contact with the CNT probe at the selected position. When the contact was made, the CNT probe would be precisely cut at the contact position (Wei et al., 2009). The cutting position can be precisely controlled using the nanomanipulators. It is found that the cutting process happened within 0.01 second by measuring the current going through the contact.

In recent years, Focused Ion Beam milling method was used to precisely shorten the CNT probes (Fang et al., 2009). The C-C chemical bonds of CNT would be broken during the high energy ion milling. The end of the CNT probe after FIB shortening is found to be a round end with fullerene-like cap, which is independent with the FIB's parameters, as shown in Fig. 5 (c). Since the FIB's beam energy is a Gaussian distribution, the Gaussian tail would produce many Carbon dangling bonds at the irradiation sites, where the dangling bonds would spontaneously close into a graphitic dome after shortening (Charlier et al., 1997).

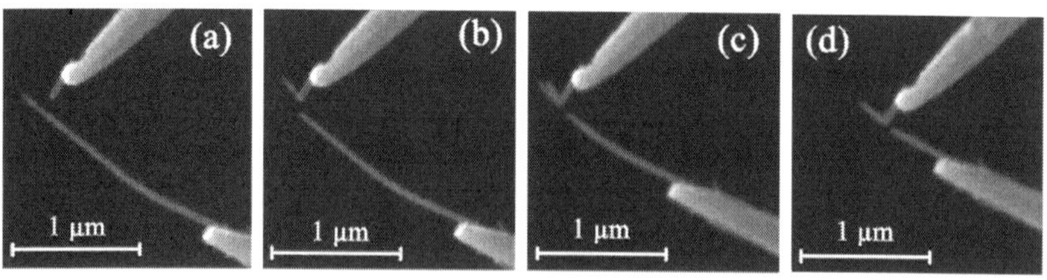

Figure 6. SEM images showing a CNT (the lower one) is cut by a CNT 'nanoknife' repeatedly and becomes shorter and shorter (Wei et al., 2007).

2.2.3. High Aspect Ratio

Shortening the CNT probe can be useful to increase its resolution and stiffness. While a short CNT probe is not always the best choice for any applications. For high aspect ratio imaging, such as grating and biological sample imaging, the CNT probe should possess sufficient aspect ratio to avoid the imaging degradation by probe broadening effect. However, the force required to bend or buckle the CNT probe would greatly decrease as its aspect-ratio increase, which would be prone to result in image artefact at the sample steep positions. A strategy to improve the mechanical integrity of CNT probe is to coat it with a thin layer film, such as polymers, Au or carbon (Yum et al., 2010).

For high aspect ratio imaging, the best CNT probe configuration should have sufficient probe lateral stiffness with portion CNT protruding at the probe end, which can get high image resolution and good probe rigidity. Method of using electron beam locally induced Pt deposition in-site to strengthen the nanotube probe with the nanotube end free of deposition was proposed (Xu et al., 2009).

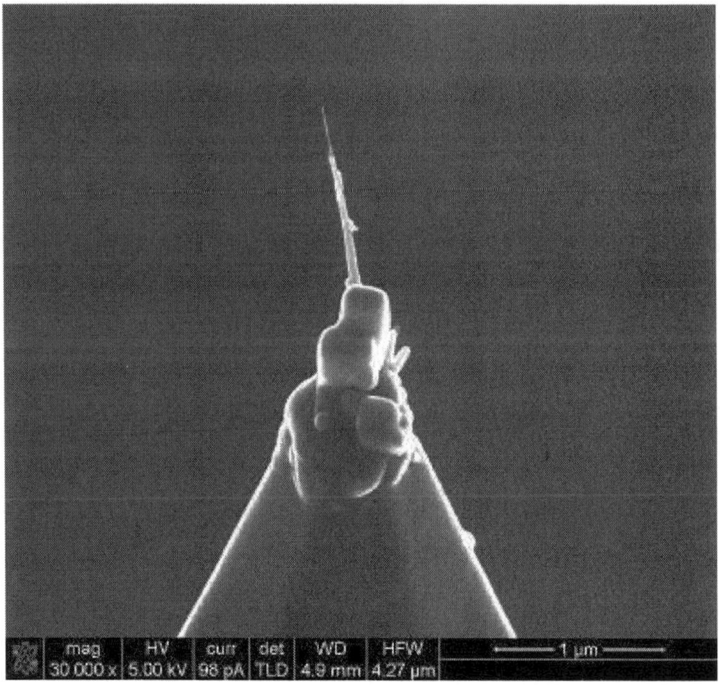

Figure 7. CNT probe for high aspect ratio imaging (Xu et al., 2009).

Besides the configurations' demands, the attachment force of CNT probe should be large enough for versatile applications under different conditions. The probe-sample interaction forces for ordinary AFM probe are typically 10–100 nN in air. The bounding force of CNT probe can be roughly calculated by measuring the Si probe's deflection before and after the CNT is cut by nanomanipulation (Xu et al., 2009). The bonding force for CNT and Si probe with Pt depostion fix method is found larger than 500nN. The direct manipulation method and CVD growth method would have sufficient bounding strength than other methods.

Although the CNT probes have been developed, but they are still not widely adopted by the AFM users. The key issue that restricts the CNT probe's wide applications is that the CNT probe is much more expensive in comparison with ordinary tapping-mode AFM probe, which is mainly due to its complex fabrication technics and less productivity. It is vital to optimize the CNT probe's fabrication process with an accessible price and meeting versatile application demands. Moreover, the CNT probe's complex mechnical response during its working would also be an obstacle for its wide applications. For example, the CNT probe would be apt to produce image artefact during the high aspect ratio image. The CNT probe-sample interaction and its mechanical response are discussed in the following section.

3. CNT PROBE-SAMPLE INTERACTION

The CNT probe is generally used in tapping-mode AFM (TM-AFM), where CNT probe intermittent contacts with sample surface and oscillating with high frequency during its scanning. Comparing with the ordinary AFM tappingmode probe, many new phenomena would appear when using a CNT probe. It is necessary to study the CNT probe interaction with sample and explain the new phenomena for CNT probe in advance.

3.1. Introductions On Tip-Sample Interaction Force And Phase Angle

Fig. 8 depicts schematically the forces acting between the AFM probe and sample as a function of the probe-sample separation D (Clemens et al., 2006). As D decreases, the AFM probe firstly detects the attractive force. If the magnitude of the attractive force gradient dF/dD exceeds the probe's force constant k (point B), then the AFM probe cantilever will be unstable and jump into contact with the sample (point B'). The position B is called "snap into contact point". As the separation further decreases, the CNT probe-sample interactions would become repulsive force.

Thus in the experiment, the ability to precisely track the force versus distance curve for all tip-to-surface separations, mainly depends on the AFM probe's force constant. For example, the unstable phenomenon of "snap to contact" would appear if the CNT probe's lateral force constant is too low. On the contrary, if the probe force constant *k* is always larger than the sample force gradient *dF/dD*, the cantilever-dependent instability can be practically eliminated, thus enable a faithful measurement of the probe-to-sample interactions (Landman et al., 1990).

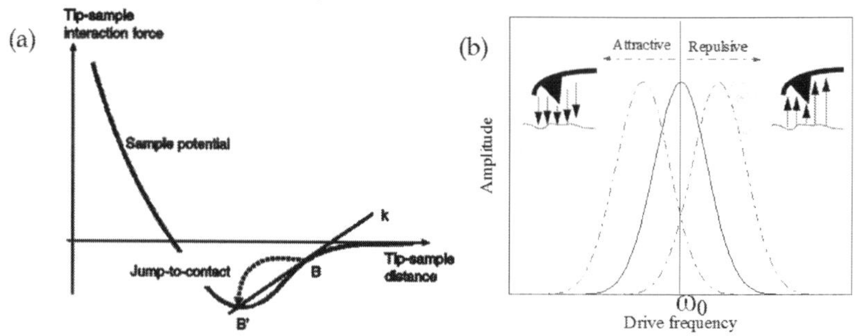

Figure 8. Interaction force curve as the tip approaches the sample (a) (Clemens et al., 2006) and the AFM probe's resonance frequency shift under different forces (b).

The AFM probe-sample interaction forces variations can be sensitively detected by the cantilever's resonance frequency shift and phase angle. Any additional tip-sample force gradient would produce a shift of the cantilever's resonant frequency for the probe in tappingmode AFM. Attractive forces make the cantilever "softer" effectively, reducing the cantilever resonant frequency. In contrast, repulsive forces make the cantilever "stiffer" effectively, increasing the resonant frequency (Fang et al., 2008).

As a function of the driving frequency ω, the phase angle φ of the cantilever oscillation relative to the signal sent to the piezoelectric driver can be derived as the following,

$$\tan\varphi = \frac{\omega\omega_0 / Q}{\omega_0^2 - \omega^2} \tag{2}$$

where the ω_0 is the probe's resonance frequency and Q is the quality constant of vibrating cantilever. From equation (2), the probe phase angle increases from 0º to 180º and the phase angle is 90º under the resonance frequency.

When the cantilever is far enough from the sample, there is no interaction between the AFM tip and the sample. Thus, the probe phase angle is 90º when the driving frequency is chosen as the cantilever resonance frequency according to equation (2). As the probe approaching the sample surface, the attractive forces first acting on the probe would decrease the probe's resonance frequency, where attractive forces make the cantilever "softer" effectively. Subsequently, the probe driving frequency becomes larger than the probe resonance frequency. Therefore, the probe phase angle is larger than 90º when the probe suffers attractive forces. Conversely, the phase angle is less than 90º when the probe works in repulsive forces region.

Therefore, the phase angle can be used to sensitively detect the probe-sample interactions. By mapping the phase of the cantilever oscillation during the AFM scan, phase imaging goes beyond simple topographical mapping to detect variations in composition, adhesion, friction, viscoelasticity, and other properties, as shown in Fig. 9. The phase images have been used to present the hydrophobic nature of single CNT by using the CNT probe with different AFM image parameters (Fang et al., 2008).

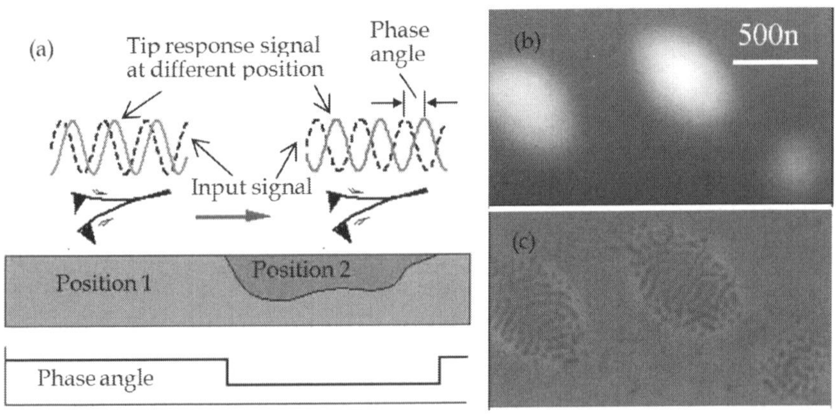

Figure 9. a) Phase angle in TM-AFM (b) topography and (c) phase images of copolymer. The height scale is 10nm and the phase angle scale is 20º (Fang et al., 2008).

3.2. IMAGE ARTEFACT

For ordinary TM-AFM Silicon probe, its force constant is large enough to avoid the instability of "snap to contact". However, in the case of CNT probe, it is very gentle in lateral directions and the CNT would elastically buckle if its lateral force exceeds the threshold value. In this section, we will discuss what the CNT probe's mechanical responses and feedback would be if the CNT probe's instability occurred. And the research works of how to overcome the instability would also present.

3.2.1. Artefact At The Steep Positions

When the CNT probe scans sample positions with a great surface slope, such as a grating step, the nanotube would be nearly parallel with the grating edge, where the lateral force exterted to the CNT probe would increase dramatically comparing with scanning the flat positions (Akita et al., 2000b). Akita has done the CNT probe force curve (force-versus-distance curve) measurement performed near the pit edge and on a plane surface, as denoted in Fig. 10. The amplitude at a plane surface monotonically decreases with the probe-sample distance, which is similar to the conventional Si probes. The CNT probe's oscillation is recovered at $z < -40$ nm due to the reproducible buckling of the nanotube. While for the case near the pit wall, the probe's oscillation is stopped abruptly, which indicates that the CNT probe snaps to contact with sample. These results from the trapping of the nanotube probe by attraction between the sidewalls of the nanotube and the pit. The adhesion force of the nanotube suffered near the pit edge is tested about 10nN (Akita et al., 2000a). At this point, it is impossible to control the position of the probe height for the AFM observation because that there is no distance dependence of the amplitude on the force curve. Therefore, the AFM image of the CNT probe at the pit edge is unstable with many ripple and wavelike distortion.

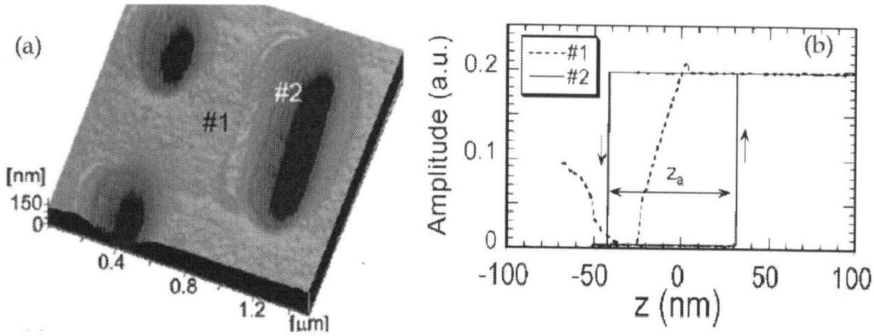

Figure 10. The CNT probe image of a DVD surface (a), and the force curve measurement performed near the pit edge and on a plane surface (b) (Akita et al., 2000a).

It was found that the ripple and wavelike distortion is more easily occurred at the pit's left edge than the right edge in general, as shown in Fig. 10. This are mainly attributed to the probe having a tilt angle of 11º when it scans a sample in AFM as shown in Fig. 3, where the vibrating probe suffers different repulsive forces at different pit edges (Fang et al., 2008).

What is the formation mechanism for the ripple and wavelike distortion at the grating edge? The phase angle data combining with height and amplitude data have been proposed to explain the image artifact, which is resulting from the CNT bent-adhesion-separation repeated process (Fang et al., 2008; Strus et al., 2005), as analyzed in Fig. 11.

When the CNT probe contacts with the grating edge, the CNT–trench wall adhesion would bend the CNT probe and then reduce the vibration amplitude of the cantilever, as position d indicates in Fig. 11. The AFM controller perceives the sudden probe amplitude drop as a surface height increase and responds by quickly moving the probe away from the sample. At position *a*, the CNT has finally been broken free from the trench wall, and the cantilever vibrates at its free amplitude. There is about a 10º phase angle decrease around position *a* in the phase angle curve at the bottom of Fig. 11, which confirms that the CNT probe really moves away from the trench wall. In order to restore the original scan amplitude, the AFM controller again lowers the probe. At position *b*, the CNT probe contacts with the trench wall again and the amplitude also drops due to the bending of the CNT. Therefore, the image artifact at the trench wall results from the CNT probe bending–adhesion–separation repeated process (Fang et al., 2008). When the ripple and wavelike AFM results appear at the steep positions using CNT probe, we should carefully judge whether it is the real sample morphology or the image artefact resulting from the CNT probe's instability.

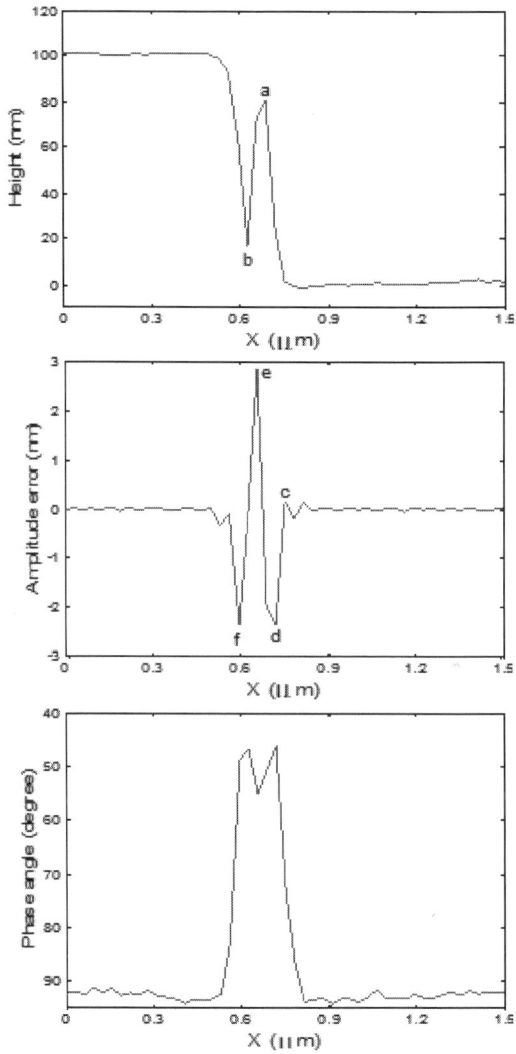

Figure 11. Sectional view of height, amplitude and phase angle of the CNT probe image artifacts (Fang et al., 2008).

The CNT probe can be treated as a series system of the ordinary AFM probe combining with a end-fixed carbon nanotube. The effective force constant for this series system can be calculated as followed:

$$K_{eff} = \frac{F}{X} = \frac{F}{\frac{F}{K_s} + \frac{F}{K_c}} = \frac{K_s K_c}{K_s + K_c} \qquad (3)$$

Therefore, the force constant of the Si probe would also affect the CNT probe's effective stiffness. Under many CNT probe fabrication experiments, the lateral force constant requirement of stable working for the CNT probe attaching to the triangular-shape AFM probe is found relative larger than the one to the rectangular shape probe. This is mainly due to the triangular shape AFM probe's less resistance ability to the lateral force (J. E. Sader & R. C. Sader, 2003).

3.2.2. Cnt-Sample Boundary Conditions

In general, CNT probes are harmonically excited near resonance and dynamically operated in either an attractive or repulsive regime using tapping-mode AFM. When CNT probes are operated in the net repulsive regime, CNTs have the potential to buckle, bend, adhere, and slide (Strus & Raman, 2009). Many studies highlighted that the CNT probe-sample boundary conditions and its mechanical responses would play a vital role in the CNT probe's stability.

A thermal noise forcing method was used to investigate the mechanical response of CNT AFM probes under the free sliding and pinned CNT contacts cases, as shown in Fig. 12. Generally, thermal noise forcing produces probe oscillation amplitudes only about angstrom. Therefore, the thermal noise spectrum is very sensitive to tiny variations of the contact properties between the CNT probe and the surface, thus providing an accurate picture of the CNT probe's mechanical response (Buchoux et al., 2009). Molecular simulations were proposed to explain the effects of CNT's slip and snap-to-contact on the resolutions (Solares & Chawla, 2010).

Strus suggested that the identification of CNT pinning and slipping during intermittent contact can be based on phase contrast imges and energy-dissipation spectroscopy, as shown in Fig. 12. The pinning or slipping of the CNT depends on the scan surface and the CNT probe's stiffness. It is found that CNT probe is easier slipping on the graphite surface than on the graphene oxide and silicon oxide surfaces (Strus & Raman, 2009).

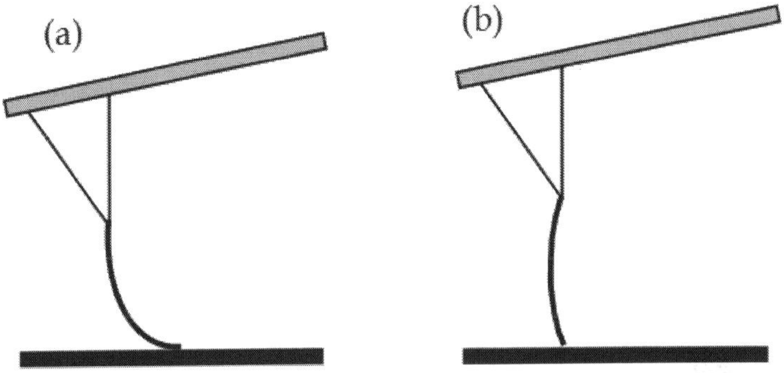

Figure 12. Illustration of the CNT probes slide (a) and pin (b) CNT contacts cases.

The CNT probe's slipping effect was experimental studied (Fang et al., 2009). It is found that the CNT probe is apt to slide if it scans with a large vibration amplitude. There are many dark lines resulting from CNT's slide in the AFM image as arrows indicated in Fig. 13(b). Due to CNT's hydrophobic property and its high aspect ratio, the capillary force between the CNT probe and the sample surface is much smaller. Therefore, the CNT probe does not need to work at large amplitudes to break the capillary force as the Si probe does and it can work well under low scanning amplitude of less than 30 nm.

The CNT slipping is an obstacle for accurate AFM topography measurements, while it can be treated as a potential method of analyzing surface composition by enhancing the probe-sample frictions contrast. AFM peeling force spectroscopy and theoretical studies have been used to quantitatively investigate the physics of adhesion and stiction of carbon nanotubes on different material substrates, such as, HOPG, polymer (Strus et al., 2008).

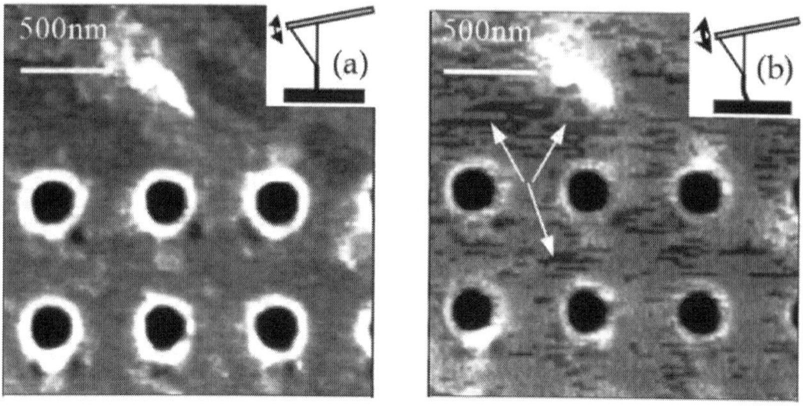

Figure 13. Nanoholes images captured by CNT probe with 18nm (a) and 40nm (b) scan amplitudes (Fang et al., 2009).

3.3. Artefact Elimination

Studies have been conducted to lessen and avoid the CNT probe's image artefact. Larger scan amplitude ratio was employed to depress the ringing artefact, by increased from 66% to 96% of the 34 nm unconstrained vibration amplitude (Strus et al., 2005). Akita found that the CNT probe's image instability near the wall of the DVD surface pits can be avoided by slightly increasing the probe's free tapping amplitude to avoid the pit's attractions (Akita et al., 2000a). Solares has theoretically explored that spectral inversion method and dual-frequency-modulation method can mitigate the imaging artifacts when using short nanotube probes (<100 nm in length) in air. The two methods are capable of performing simultaneous imaging and force spectroscopy (Solares & Chawla, 2010).

The most effective method to overcome the CNT probe's image artefact is to strengthen the CNT probe, as discussed in the Section 2.2. For example, several micrometers long multiwalled CNT probes were prepared and strengthened with carbon molecular layers to overcome mechanical instabilities (Vakarelski et al., 2007).

4. CNT PROBE'S APPLICATIONS

4.1. High Resolution Imaging

CNT's excellent properties make CNT probe be a favorable choice for AFM probe in topographic imaging. SWNT probe with several nanometers' radius shows great image resolution over conventional Si probe (Cheung et al., 2000). CNT probes have also been showed higher image resolution for the high aspect ratio samples, such as, grating, cellular samples, etc (Dai et al., 1996;Fang et al., 2007; Hafner, 1999, 2001; Stevens et al., 2000). It is found that not only the CNT probe's radius would influence its resolution, its end configuration also showed impact on CNT probe's image ability (Fang et al., 2009).

CNT has a much higher Young's modulus and exceptional elastic buckling property, which results in the much better wear resistance ability of the CNT probes relative to the Si probes. The excellent wear resistance property makes CNT probes have consistent image resolution, which is very important for applications, such as, online inspection.

Electrostatic force microscopy (EFM) is one technique sensitive to the long-range electrostatic forces, which are useful in the study of semiconductor devices. Because of the long-range nature of electrostatic force, interpretation of EFM images is often complicated as the probe beyond the probe apex would also affect the EFM image results. This effect can be greatly reduced using high aspect ratio CNT probes (Wilson & Macpherson, 2009). For similar reasons CNT probes are also suitable for high resolution magnetic force microscopy. Iron-filled carbon nanotubes (Fe-CNTs) have been proposed for the application for magnetic force microscopy (Wolny et al., 2008).

4.2. Nanofabrication

Carbon nanotube probe is not only a tool for imaging, but also for nanofabrication, as shown in Fig. 14. AFM nanolithography can be widely applied in fields such as data storage and device fabrication. Due to nanotube's excellent properties, carbon nanotube probes have been explored in fabricating oxide nanostructures on silicon surfaces (Dai & Franklin, 1998), a p-GaAs(100) surface (Huang et al., 2006), a metallic tantalum film (Choi et al., 2007).

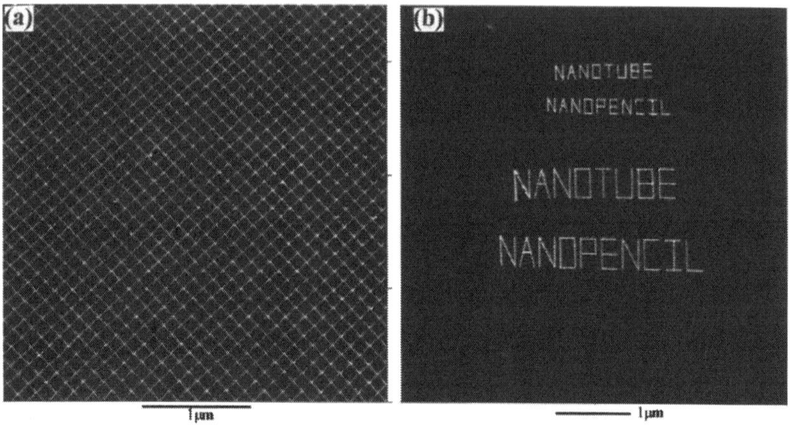

Figure 14. AFM image of 2 nm tall, 10 mm wide, and 100 nm spaced silicon-oxide (light) lines fabricated by a nanotube tip using anodic oxidation method (Dai & Franklin, 1998).

It was found that hydrophobicity is a key factor in the improved reliability of CNT probes on the nanoscale oxide fabrication over conventional Si probes (Kuramochi et al., 2007). Okazaki studied the nano-gratings direct nanolithography on organic polysilane PMPS films using CNT probe (Okazaki et al., 2000).

4.3. Biology Application

Due to its excellent properties, CNT probe has been widely used in the biological application. It produces lower damage to the proteins than the Si probe during AFM scanning, as shown in Fig. 15(Fang et al., 2007). CNT probe has been proven more suitable for imaging soft biological samples, such as, DNA, protein (Cheung et al., 2000).

Technology for introducing biomolecules and nanoparticles into living cells with minimal invasiveness is a key task to study the physical properties and biochemical interactions that govern the cell's behavior (Chen et al., 2007). Micro glass pipettes have been used for intracellular analysis research. However, due to their relatively large sizes and rigid structure, these probes cause severe damage to the cells (Niu et al., 2011).

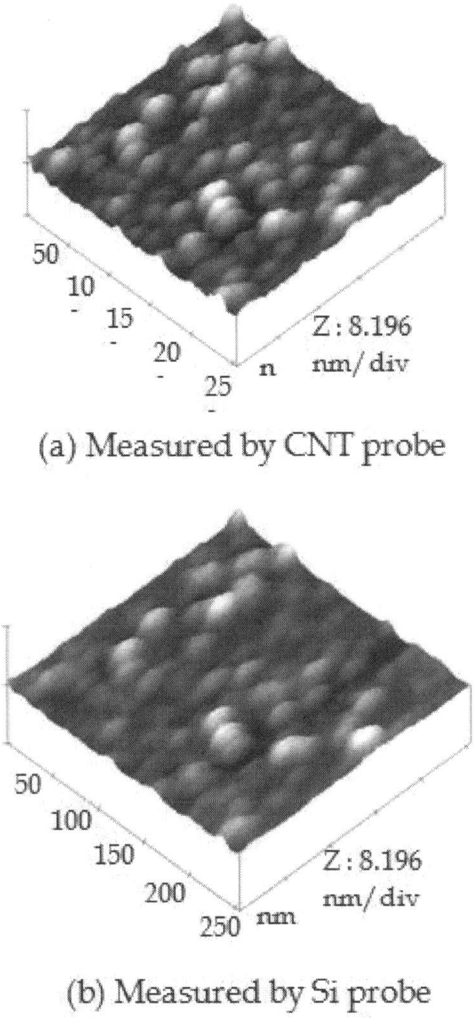

Figure 15. AFM analyses of IgG protein. The Si probe has created a larger probe indentation to the soft proteins than the CNT probe (Fang et al., 2007).

Due to its nanoscopic dimensions, high mechanical strength, and functionalizable surfaces, CNT probe has been used as a "nanoneedle" in a nanoscale cell injection system to deliver cargo into cells. A multiwalled CNT probe was functionalized with cargo via a disulfide-based linker. Penetration of cell membranes with this "nanoneedle" was controlled by the AFM. The following reductive cleavage of the disulfide bonds within the cell's interior resulted in the release of cargo inside the cells, after which the nanoneedle was retracted by the AFM control, as shown in Fig. 16 (Chen et al., 2007). The CNT probe "nanoneedle" technique causes little membrane or cell damage. Other biomolecules such as DNA and RNA, or synthetic structures such as polymers and nanoparticles can be delivered into cells in a similar fashion (Chen et al., 2007; Vakarelski et al., 2007).

Gold nanoparticle-decorated CNT probes are used to study intracellular environments in situ using surface-enhanced Raman spectroscopy (SERS), as shown in Fig. 17 (Niu et al., 2011). The high aspect ratio of CNTs displaces less intracellular volume than ordinary glass pipettes or optical fibers cell probes, thus perturbing the cell much less, especially when measurements performed over long periods of time. This might be a useful method for tip-enhanced Raman spectroscopy (TERS) application using CNT probe.

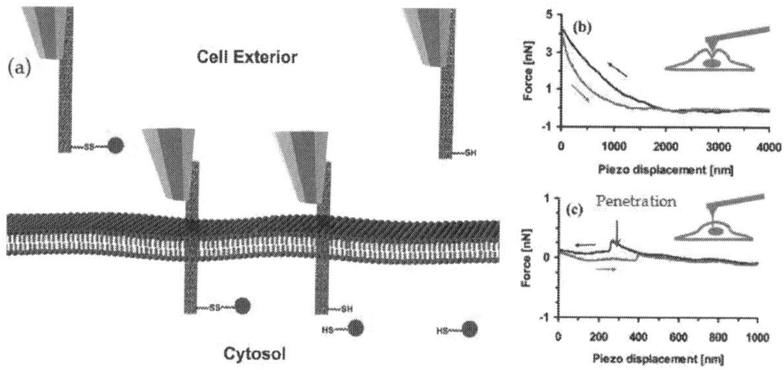

Figure 16. Schematic of the CNT probe nanoinjection procedure. A CNT probe with cargo attached to the CNT surface via a disulfide linker penetrates a cell membrane (a) (Chen et al., 2007). Force vs piezo displacement curves for the indentation of a living cell using a pyramidal Si probe (b) and a C/Au-coated nanotube probe (c) (Vakarelski et al., 2007).

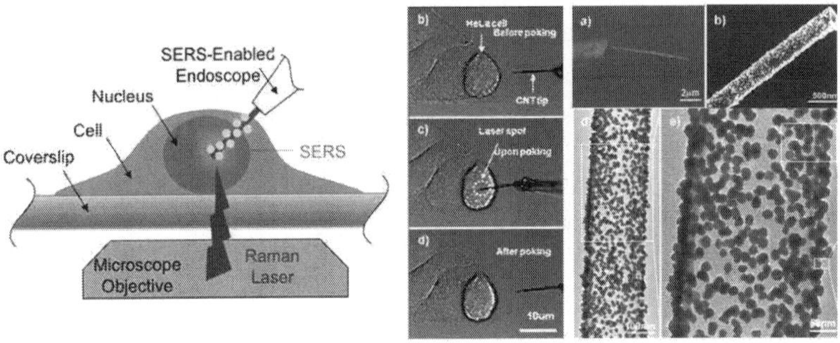

Figure 17. Intracellular SERS study with a nanoparticle-decorated CNT probe (Niu et al., 2011). (a) Intracellular SERS study with a CNT probe (b) SEM of gold-decorated CNT probe

5. SUMMARY AND OUTLOOK

Carbon nanotube probes open up a new AFM imaging world, increasing the probe's resolution and longevity, decreasing probe–sample forces, and extending the AFM application fields. It would also have a significant impact in key research areas, such as, nanometrology, surface engineering and biotechnology.

Future main attentions in the carbon nanotube AFM probes can be summarized as follows:

- CNT probe fabrication technics' optimization: the technics of high efficient fabrication with low cost; the fabrication's reproducibility and controllability.
- CNT probe imaging accuracy calibration: the CNT-sample boundary conditions and CNT's mechanical response effect studies on the CNT probe imaging resolutions in nanometer or angstrom levels; CNT probe's accuracy comparison with the others precision measurement techniques.
- CNT probe's unique applications: TERS; the biological application, such as, functionalized CNT probe as nanoneedle for intracellular studies and drug delivery.

6. ACKNOWLEDGEMENTS

The authors would thank the persons who have given kind helps in our CNT probe's researches, Dr. L. Q. Guo from Harbin Institute of Technology, Prof. Jason H. Hafner from Rice University, Prof. Hongjie Dai from Stanford University, Dr. Jason Cleveland from Asylum Research and Dr. M. C. Strus from Purdue University. Thanks go for Mr. Haifeng Gao and Miss Lihua Xu from Tianjin University for helping the typesetting.

The research work is supported by '111' project by the State Administration of Foreign Experts Affairs and the Ministry of Education of China (B07014), National High Technology Research and Development Program of China (863 No. 2009AA044305& 2009AA044204), National Natural Science Foundation of China (No. 50905126, 90923038&50935001), Ministry of Industry and Information Technology (No. 2011ZX04014-071), Natural Science Foundation of Tianjin(No. 10JCYBJC06400), and National Basic Research Program of China (973 Program, Grant No.2011CB706700).

REFERENCES

1. S. Akita, H. Nishijima, Y. Nakayama, 2000a Influence of stiffness of carbon-nanotube probes in atomic force microscopy. J. Phys. D: Appl. Phys., 33 21 (November 2000), 26732677 , 0022-3727

2. S. Akita, H. Nishijima, T. Kishida, Y. Nakayama, 2000b Influence of Force Acting on Side Face of Carbon Nanotube in Atomic Force Microscopy. Jpn. J. Appl. Phys., 39 (March 2000), 37243727 , 0021-4922

3. G. Binnig, C. Quate, C. Gerber, 1986 Atomic Force Microscope. Phys. Rev. Lett., 56 9 (March 1986), 930 933, 0031-9007

4. J. Buchoux, J. P. Aim´e, R. Boisgard, C. V. Nguyen, L. Buchaillot, S. Marsaudon, 2009 Investigation of the carbon nanotube AFM tip contacts: free sliding versus pinned contact. Nanotechnology, 20 47 (November 2009), 475701 (8pp), 0957-4484

5. J. C. Charlier, A. D. Vita, X. Blasé, R. Car, 1997 Microscopic Growth Mechanisms for Carbon Nanotubes. Science, 275 (January 1997), 647649 , 0036-8075

6. X. Chen, A. Kis, A. Zettl, C. R. Bertozzi, 2007 A cell nanoinjector based on carbon nanotubes. PNAS, 104 20 (May 2007), 82188222

7. C. L. Cheung, J. H. Hafner, C. M. Lieber, 2000 Carbon nanotube atomic force microscopy tips: Direct growth by chemical vapor deposition and application to high-resolution imaging. Proc. Natl Acad. Sci. USA, 97 8 (April 2000), 38093813 , 0027-8424

8. J. S. Choi, S. Bae, S. J. Ahn, D. H. Kim, K. Y. Jung, C. Han, C. C. Chung, H. Lee, 2007 Feasibility of multi-walled carbon nanotube probes in AFM anodization lithography. Ultramicroscopy, 107 10-11 , (October 2007), 10911094 , 0304-3991

9. M. F. Clemens, D. G. Sunil, S. Kim, S. G. Kim, 2006 A Nanoscanning Platform for Bio-Engineering: an in-Plane Probe with Switchable Stiffness. Nanotechnology, 17 4 (February 2006), S69 S76, 0957-4484

10. H. J. Dai, J. H. Hafner, A. G. Rinzler, D. T. Colbert, R. E. Smalley, 1996 Nanotubes as Nanoprobes in Scanning Probe Microscopy. Nature, 384 (November 1996), 147 151, 1476-1122

11. H. J. Dai, N. Franklin, 1998 Exploiting the Properties of Carbon Nanotubes for Nanolithography. Applied Physics Letters, 73 11 (September 1998), 1508 1510, 0003-6951

12. Z. F. Deng, E. Yenilmez, A. Reilein, J. Leu, H. J. Dai, K. A. Moler, 2006 Nanotube Manipulation with Focused Ion Beam. Applied Physics Letters, 88 2 (January 2006), 023119 (3 pp), 0003-6951

13. J. P. Edgeworth, D. P. Burt, P. S. Dobson, J. M. R. Weaver, J. V. Macpherson, 2010 Growth and morphology control of carbon nanotubes at the apexes of pyramidal silicon tips. Nanotechnology, 21 10 (March 2010), 105605 (10pp), 0957-4484

14. F. Z. Fang, Z. W. Xu, S. Dong, G. X. Zhang, 2007 High aspect ratio nanometrology using carbon nanotube probes in atomic force microscopy. Annals of the CIRP, 56 1 (2007), 533 536, 0007-8506

15. F. Z. Fang, Z. W. Xu, S. Dong, 2008 Study on Phase Images of Carbon Nanotube Probe in Atomic Force Microscopy. Measurement Science and Technology, 19 5 (May 2008), 055501 7 pp), 0957-0233

16. F. Z. Fang, Z. W. Xu, G. X. Zhang, X. T. Hu, 2009 Fabrication and configuration of carbon nanotube probes in atomic force microscopy. Annals of the CIRP, 58 1 (April 2009), 455 458, 0007-8506

17. J. H. Hafner, C. L. Cheung, C. M. Lieber, 1999 Growth of Nanotubes for Probe Microscopy Tips. Nature, 398 6730 (April 1999), 761 762, 1476-1122

18. J. H. Hafner, C. L. Cheung, T. H. Oosterkamp, C. M. Lieber, 2001 High-Yield Assembly of Individual Single-Walled Carbon Nanotube Tips for Scanning Probe Microscopies. Journal of Physical Chemistry B, 105 4 (February 2001), 743 746, 1520-6106

19. A. Hall, W. G. Matthews, R. Superfine, M. R. Falvo, S. Washburn, 2003 Simple and Efficient Method for Carbon Nanotube Attachment to Scanning Probes and Other Substrates. Appl. Phys. Lett., 82 15 (April 2003), 2506 2508, 0003-6951

20. W. P. Huang, H. H. Cheng, S. R. Jian, D. S. Chuu, J. Y. Hsieh, C. M. Lin, M. S. Chiang, 2006 Localized electrochemical oxidation of p-GaAs(100) using atomic force microscopy with a carbon nanotube probe. Nanotechnology, 17 15 (August 2006), 38383843 , 0957-4484

21. S. Iijima, 1991 Helical Microtubules of Graphitic Carbon. Nature, 354 (November 1991), 56 58, 1476-1122

22. S. Kim, H. W. Lee, S. G. Kim, 2009 Transplanting assembly of carbon-nanotube-tipped atomic force microscope probes. Applied Physics Letters, 94 19 (May 2009), 193102, 0003-6951

23. A. V. Krasheninnikov, K. Nordlund, J. Keinonen, 2002 Production of defects in supported carbon nanotubes under ion irradiation. Phys. Rev. B, 65 16 (April 2002), 165423, 1098-0121

24. A. V. Krasheninnikov, K. Nordlund, 2010 Ion and electron irradiation-induced effects in nanostructured materials. J. Appl. Phys., 107 7 (April 2010), 071301, 0021-8979

25. H. Kuramochi, T. Tokizaki, H. Yokoyama, J. A. Dagata, 2007 Why nano-oxidation with carbon nanotube probes is so stable: I. Linkage between hydrophobicity and stability. Nanotechnology, 18 13 (April 2007), 135703 (6pp), 0957-4484

26. W. C. Lai, S. C. Chin, Y. C. Chang, L. Y. Chen, C. S. Chang, 2010 Lattice-resolved frictional pattern probed by tailored carbon nanotubes. Nanotechnology, 21 5 (February 2010), 055702 (9pp), 0957-4484

27. U. Landman, W. D. Luedtke, N. A. Bumham, R. J. Colton, 1990 Atomistic mechanisms and dynamics of adhesion, nanoindentation, and fracture. Science, 248 4594 (April 1990), 454611 , 0036-8075

28. H. Nishijima, S. Kamo, S. Akita, Y. Nakayama, 1999 Carbon-Nanotube Tips for Scanning Probe Microscopy: Preparation by a Controlled Process and Observation of Deoxyribonucleic Acid. Appl. Phys. Lett, 74 26 (June 1999), 4061 4063, 0003-6951

29. J. J. Niu, M. G. Schrlau, G. Friedman, Y. Gogotsi, 2011 Carbon Nanotube-Tipped Endoscope for In Situ Intracellular Surface-Enhanced Raman Spectroscopy. Small, 7 4 (February 2011), 540545 , 1613-6810

30. A. Okazaki, T. Kishida, S. Akita, H. Nishijima, Y. Nakayama, 2000 Direct Nanolithography of Organic Polysilane Films Using Carbon Nanotube Tips. Jpn. J. Appl. Phys., 39 12B (December 2000), 7067 7069, 0021-4922

31. B. C. Park, K. Y. Jung, W. Y. Song, O. Beom-Hoan, S. J. Ahn, 2006 Bending of a carbon nanotube in vacuum using a focused ion beam, Advanced Materials, 18 1 (January 2006), 9598 , 0935-9648

32. J. E. Sader, R. C. Sader, 2000 Susceptibility of Atomic Force Microscope Cantilevers to Lateral Forces: Experimental Verification. Appl. Phys. Lett., 83 15 (October 2000), 3195 3197, 0003-6951

33. S. D. Solares, G. Chawla, 2010 Exploration of AFM Imaging Artifacts Occurring at Sharp Surface Features When Using Short Carbon Nanotube Probes and Possible Mitigation With Real-Time Force Spectroscopy. Journal of Manufacturing Science and Engineering, 132 3 (June 2010), 030904(14pp), 1087-1357

34. R. M. D. Stevens, N. A. Frederick, B. L. Smith, D. E. Morse, G. D. Stucky, P. K. Hansma, 2000 Carbon Nanotubes as Probes for Atomic Force Microscopy. Nanotechnology. 11 1 (March 2000), 1 5, 0957-4484

35. M. C. Strus, A. Raman, C. S. Han, C. V. Nguyen, 2005 Imaging artefacts in atomic force microscopy with carbon nanotube tips. Nanotechnology, 16 11 (November 2005), 24822492 , 0957-4484

36. M. C. Strus, L. Zalamea, A. Raman, R. B. Pipes, C. V. Nguyen, E. A. Stach, 2008 Peeling Force Spectroscopy: Exposing the Adhesive Nanomechanics of One-Dimensional Nanostructures. Nano Letters, 8 2 (February 2008), 544550 , 1530-6984

37. M. C. Strus, A. Raman, 2009 Identification of multiple oscillation states of carbon nanotube tipped cantilevers interacting with surfaces in dynamic atomic force microscopy. Physical Review B, 80 22 (December 2009), 224105, 1098-0121

38. S. D. Tamara, H. Stephanie, S. S. Ulrich, 2010 Microwave-Assisted Fabrication of Carbon Nanotube AFM Tips. Nano Lett., 10 10 (October 2010), 40094012 , 1530-6984

39. J. Tang, G. Yang, Q. Zhang, A. Parhat, B. Maynor, J. Liu, L. C. Qin, O. Zhou, 2005 Rapid and Reproducible Fabrication of Carbon Nanotube AFM Probes by Dielectrophoresis. Nano Letters, 5 1 (January 2005), 11 14, 1530-6984

40. I. U. Vakarelski, S. C. Brown, K. Higashitani, Moudgil, B. M. , 2007 Penetration of Living Cell Membranes with Fortified Carbon Nanotube Tips. Langmuir, 23 22 (October 2007), 1089310896 , 0743-7463

41. X. L. Wei, Q. Chen, Y. Liu, L. M. Peng, 2007 Cutting and Sharpening Carbon Nanotubes Using a Carbon Nanotube 'Nanoknife'. Nanotechnology, 18 18 (May 2007), 185503, 0957-4484

42. X. L. Wei, A. N. Jiang, S. Gao, Q. Chen, 2009 Length Control and Sharpening of Carbon Nanotube Scanning Probe Microscope Tips Using Carbon Nanotube "Nanoknife". J. Nanosci. and Nanotechnol., 9 2 (February 2009), 12581262 , 1533-4880

43. N. R. Wilson, J. V. Macpherson, 2009 Carbon Nanotube Tips for Atomic Force Microscopy. Nature nanotechnology, 4 8 (August 2009), 483 491, 0957-4484

44. F. Wolny, U. Weissker, T. Mühl, A. Leonhardt, S. Menzel, A. Winkler, B. Büchner, 2008 Iron-filled carbon nanotubes as

probes for magnetic force microscopy. Journal of Applied Physics, 104 6 (September 2008), 064908, 0021-8979

45. Z. W. Xu, F. Z. Fang, X. T. Hu, 2009 Fabrication of Carbon Nanotube Probes in Atomic Force Microscopy. Advanced Materials Research, 76-78 , (June 2009), 497501 , 1022-6680

46. Q. Ye, A. M. Cassell, H. B. Liu, K. J. Chao, J. Han, M. Meyyappan, 2004 Large-Scale Fabrication of Carbon Nanotube Probe Tips for Atomic Force Microscopy Critical Dimension Imaging Applications. Nano Letters, 4 7 (July 2004), 1301 1308, 1530-6984

47. K. Yum, N. Wang, M. F. Yu, 2010 Nanoneedle: A multifunctional tool for biological studies in living cells. Nanoscale, 2 3 (March 2010), 363372 , 2040-3372

4

FREQUENCY NOISE PROPERTIES OF LASERS FOR INTERFEROMETRY IN NANOMETROLOGY

Jan Hrabina *, Josef Lazar, Miroslava Holá and Ondřej Číp

Institute of Scientific Instruments, v.v.i., Academy of Sciences of the Czech Republic, Královopolská 147, Brno 61264, Czech Republic

ABSTRACT

In this contribution we focus on laser frequency noise properties and their influence on the interferometric displacement measurements. A setup for measurement of laser frequency noise is proposed and tested together with simultaneous measurement of fluctuations in displacement in the Michelson interferometer. Several laser sources, including traditional He-Ne and solid-state lasers, and their noise properties are evaluated and compared. The contribution of the laser frequency noise to the displacement measurement is discussed in the context of other sources of uncertainty associated with the interferometric setup, such as, mechanics, resolution of analog-to-digital conversion, frequency bandwidth of the detection chain, and variations of the refractive index of air.

Keywords: nanometrology; laser noise; interferometry; nanopositioning; AFM

1. INTRODUCTION

Laser interferometry is the fundamental measuring technique for length and all dimensional quantities in fundamental metrology as well as in demanding industrial applications. The measurement range of laser interferometry covers the nano-world in the most precise devices like metrological scanning probe microscopes (SPMs), middle-sized objects, measured via coordinate measurement machines (CMMs), and up to kilometer distances in long-distance measurements [1–5]. The design of interferometric systems needs a great care when demands for precision are high. The whole arrangement of a laser interferometer consists of several key components—a laser source which powers the optical section of the interferometer, generating precise wavelength acting as a "ruler scale" in distance measurement, optics and mechanics designed according to the application, and finally the detection chain for acquisition, processing and evaluation of the interference signal including electronic hardware and software (demodulation and linearization techniques) [6–14]. Each component plays a role in the evaluation of the overall uncertainty.

1.1. Laser Source Noise Properties

As the laser interferometer relies on coherent laser light and actually counts discrete wavelengths, the intensity and mainly frequency noise of the laser source can significantly influence the uncertainty of the interferometric measurement. These properties have to be taken into consideration—especially frequency noise of the laser is directly transferred into the noise of the measured distance (quantity).

1.1.1. Slow Frequency Noise—Drift

The most common type of lasers used in practical metrology of length are the relatively cheap He-Ne lasers operating at 633 nm wavelength. They offer relatively good long-term frequency stability up to range of 10^{-11} when frequency-stabilized by saturated absorption of molecular iodine [15–17]. Stabilization to the active line in Ne offers stability at the 10^{-8} level which is fully sufficient for measurements done under atmospheric conditions [18–21]. The main limitation in this case are the fluctuations of the refractive index of air (which can be controlled down to the range of approx. 10^{-7}) [22–24].

The main disadvantage of He-Ne lasers is their limited power. Especially in multidimensional systems like CMM machines, where more interferometers are supplied from a single laser source, the He-Ne laser power can be a limiting factor. Over several years frequency doubled Nd:YAG lasers have become popular in metrology of length. They offer more power able to feed many measuring axes, a slightly shorter wavelength of 532 nm means better resolution and also better noise properties can be observed. The long-term frequency stability of the Nd:YAG lasers close to the 10^{-14} level for an integration time of 1,000 s can be achieved in optical setups where saturation spectroscopy in molecular iodine techniques are implemented [25–32]. In less demanding setups (interferometry under atmospheric conditions) for example, simple linear spectroscopy in molecular iodine can do the job to reduce long-term drift and offers frequency stability of 5×10^{-9} for 100 s [33–38].

1.1.2. Frequency Noise

Faster frequency fluctuations of the laser source can degrade the interferometric measurement [39,40]. Especially in applications that cover nanometrology, where the demands are the highest and high-speed scanning and, thus large bandwidth are needed. This is why in this contribution we have focused especially on the measurement of frequency noise of various laser sources intended for use in laser interferometry. The laser noise analysis will be described in the experimental section together with description of the experimental setup. The problem of laser noise is related also to the frequency bandwidth, other properties of the detection system and length of the measuring arm of the interferometer.

1.2. Interferometer Configuration

Any interferometric arrangement with reference and measuring beam paths is sensitive to the wavelength of the laser source in relation to the difference in their optical lengths. Low-coherent or white-light interferometry exploits this effect. The center of a white-light fringe gives the information about balance between the two optical paths [41,42]. This is valid also for Michelson interferometer designed for measurement of displacement with a highly coherent laser source subject to optical frequency noise. Variations of the laser optical frequency seen either as a long-term drift (instability) or a frequency (phase) noise contribute to the measurement uncertainty more, when there is a larger difference between the measuring and reference optical lengths and do not contribute at all when the arms are precisely balanced [5,40,43].

A similar effect in nature is caused by fluctuations of the refractive index of air. This affects the speed of light in atmospheric conditions and thus the conversion of optical frequency into wavelength, which is actually what the interferometer counts. The value of refractive index can be evaluated from the parameters of atmosphere down to range of 10^{-7}[22–24] or more or less effectively compensated [44–50].

1.3. Resolution of A/D Converters and Frequency Bandwidth

From the point of view of the interference signal processing there comes the resolution of analog-to-digital converters in the detection chain. An interference fringe signal is sampled with this resolution and resolution of A/D converters is sometimes interpreted as the resolution of the interferometer. A dependency between A/D resolution and simple Michelson interferometer resolution is in Table 1 for 532 and 633 nm wavelength respectively. The resolution (minimal detectable position change of the measuring mirror) can be evaluated by Equation (1):

Table 1. Dependency between the resolution of A/D conversion and resolution (minimal detectable distance) of the interferometer for the 532 and 633 nm wavelengths respectively.

	Resolution of the interferometer [nm]	
A/D resolution	$\lambda = 532$ nm	$\lambda = 633$ nm
12 bits	6.49	7.73
14 bits	1.62	1.93
16 bits	0.41	0.48
18 bits	0.10	0.12
20 bits	0.025	0.030
22 bits	0.0063	0.0076
24 bits	0.0016	0.0019

$$L_{MIN} = \frac{\lambda}{2^x \times n} \tag{1}$$

where λ is wavelength of incident light, x is bit-resolution of the analog-to-digital converter and finally n is the number of measuring beams passing through the measured path in a multipass configuration (in case of setup like at Figure 1 n = 2). The resolution of the interferometer is influenced by the laser and other sources of noise and interference and has to be considered together with the desired frequency (speed) bandwidth of the system. Multipass setup may be an option.

This approach may lead to unrealistic expectations. For example for 18-bit resolution of A/D converter we could obtain 1 pm displacement resolution of the interferometer, but when there is a difference in lengths of reference and measuring arms, frequency fluctuation above this limit will also influence the result (not to speak about the refractive index), so the frequency/phase noise effects have to be considered. Next, especially in case of higher bit resolution of the A/D converter other sources of the noise start to influence the measurement and should be considered and precisely controlled. For example noise of the used electronics, noise floor of the photodetectors and proper electrical grounding and shielding should be taken into account. These properties are mainly of technical character and their detail analysis depends on concrete solution of the interferometric setup.

The frequency bandwidth of the whole detection chain plays also a role. This parameter is closely related to the spectral properties of the laser source and every source of variations (including fluctuations of refractive index of air, acoustic noise, *etc.*) has to be judged from the point of view of their speed and the required speed of the measurement.

From the point of view of the overall uncertainty of the measurement also interferometer nonlinearity errors should be addressed and taken into account. The nonlinearity of the phase detection can be effectively suppressed by proper linearization methods published before [8–11,13,14,43].

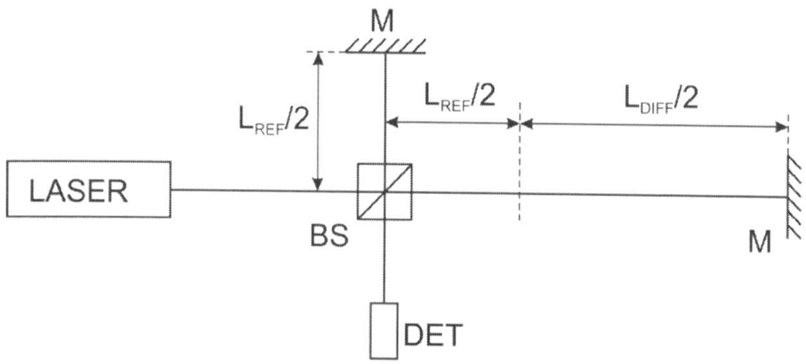

Figure 1. Principal schematic of a Michelson interferometer; BS—beamsplitter, M—mirrors, DET—photodetector, L_{REF}—optical length of reference arm, L_{DIFF}—difference distance between optical lenghts of reference and measuring arm of the interferometer.

2. ANALYSIS OF THE INTERFEROMETER

Considering the Michelson interferometer optical setup as in Figure 1, from the point of view of the frequency noise and frequency fluctuations the key parameter is the length of L_{DIFF}—difference between optical lengths of reference and measuring arm of the interferometer. The influence of the frequency noise level will be higher in case of higher optical length difference between both arms. When this contribution reaches the level of $\lambda/2$ the interferometer loses the count of fringes—the interferometer goes beyond the coherence length of the laser. This also means that interferometers with arms of similar lengths (for example for measuring within a short range) will be less sensitive to the frequency fluctuations of the laser source.

When we want to define a sensitivity of the interferometer to the frequency noise of the laser, we can use an expression of minimal frequency change which will influence the result of measurement (or which will exceed the quantization noise/resolution of the A/D converter). A length change ΔL seen at the output of the system caused by frequency shift of the laser light can be expressed as Equation (2):

$$\Delta L = L_{DIFF} \times \left(\frac{f_0}{f_{ACT}} - 1 \right) \qquad (2)$$

where L_{DIFF} is distance difference between reference and measuring arms, f_{ACT} is actual (shifted) optical frequency and f_0 is the central optical frequency of the laser source ($f_0 = c/\lambda$). A dependency between the length difference between the interferometer arms and a limiting level of frequency noise exceeding the quantization noise is in Figure 2 for different resolutions of the A/D conversion and the wavelength 532 nm of the laser.

Next aspect related with the laser source frequency noise properties which should be considered is the frequency bandwidth of the system. A maximal speed of displacement of the interferometer measuring path v_{MAX} which can be detected by the system from Figure 1 can be simply described as in Equation (3):

$$v_{MAX} = \frac{\lambda \times f_{SA}}{4} \qquad (3)$$

where λ is wavelength of incident light and f_{SA} is sampling frequency of the analog-digital converter. The example values of these reacheable moving speeds for 532 and 633 nm wavelengths and three different sampling speeds are shown in Table 2.

In this contribution we concentrate on one of the sources of uncertainty in interferometry related to the laser noise. Considering facts described above we decided to investigate the short-term amplitude and also frequency noise properties of several different laser sources and tried to compare and evaluate their properties.

Table 2. Maximal detectable speeds of distance changes *vs.* f_{SA} of ADC.

f_{SA} [kHz]	v_{max} [m/s], λ = 532 nm	v_{max} [m/s], λ = 633 nm
0.1	2.66e-5	3.16e-5
1	2.66e-4	3.16e-4
10	2.66e-3	3.16e-3
100	2.66e-2	3.16e-2
300	7.98e-2	9.48e-2

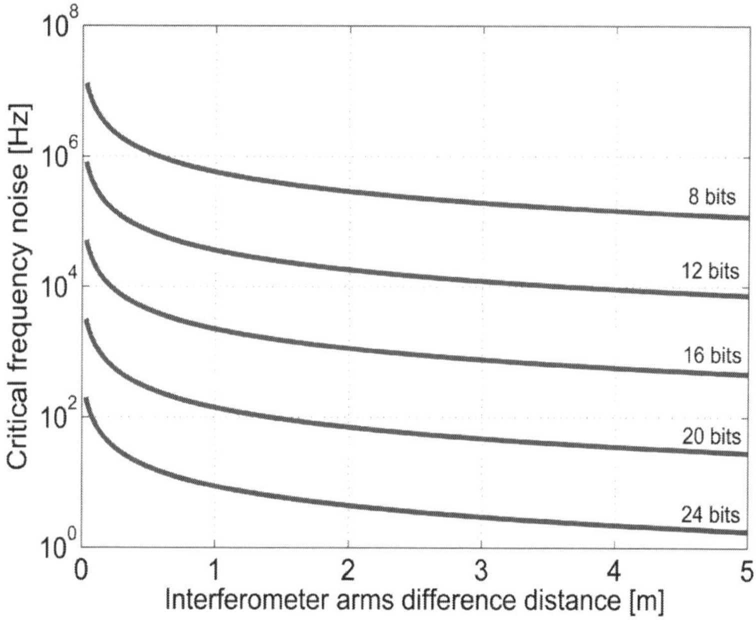

Figure 2. Length difference between the measuring and reference arms of the interferometer and corresponding limiting level of frequency fluctuations for 532 nm wavelength. (resolutions of the A/D conversion from top: 8, 12, 16, 20, 24 bits).

3. EXPERIMENTAL SECTION

We set up an experimental arrangement for simultaneous measuring of both amplitude and short-term frequency noise of several single-frequency He-Ne lasers and also frequency doubled Nd:YAG lasers. The optical power amplitude fluctuations were measured directly by a photodetector. Frequency noise of the laser radiation was measured with the help of a passive Fabry-Perot cavity used as an optical frequency discriminator. This cavity contained a mirror holder equipped with piezoelectric element which allowed tuning of the cavity length. The length of tunable Fabry-Perot cavity was in the next step stabilized by slow servo-loop ($\tau \sim 3$ s) to the value of investigated laser optical frequency so the frequency of the laser matched the middle point of the slope of the resonant transmission curve of the cavity. In this regime the cavity operated as a frequency discriminator suitable for fast laser noise fluctuations measurement. The free spectral range of the cavity was 2 GHz and measured linewidth of the cavity was 25 MHz. While the output signal of this frequency discriminator contained both the amplitude and frequency noises, the amplitude noise measured directly by photodetector and frequency fluctuations investigated through the cavity were recorded simultaneously. This gave us an opportunity to subtract the influence of amplitude noise from the frequency discriminator output. The cavity was inserted into an evacuated and thermal-shielded chamber to suppress the influences of refractive index of air fluctuations.

The correct operation of the frequency discriminator was evaluated by the comparison of results from second frequency noise measurement through the reference Michelson interferometer with homodyne quadrature detection. The interferometer reacts to variations of the input laser frequency by variations of the phase of the signal on the output of the quadrature detection unit so it can be seen as another instrument of a frequency discriminator operating within the range of a single interference fringe. The key advantage of this approach is only negligible influence of amplitude fluctuations of the laser power to the output signal because the detection system response to these fluctuations is a few times smaller in comparison to frequency noise. We used a Michelson interferometer in four-pass configuration and a flat mirror reflector. The optical length difference between reference and measurement arm of the interferometer was 2 m (0.5 m mechanical distance) and the measurement were performed in thermal-shielded box to minimize the influence of the environmental conditions. Both measurements (with the help of the Fabry-Perot cavity and also with the help of interferometer) were performed in the same time to have a possibility to compare the results. One of the tested laser was equipped with a resonator mirror holder supplemented with piezoceramic element (PZT). This option enable us to frequency modulate this laser throught the voltage change of this PZT. This PZT was modulated by sinusoidal signals of different frequencies (1–30 kHz) and amplitudes to have a possibility to separate the modulation influence from the other frequency noise sources and results from cavity and also interferometer measurements were compared (frequency widths of this modulation were selected few orders higher than the noise background of the system itself). Results show very good correlation between 1st harmonics of the modulation signal. Measured signals were simultaneously recorded with the digital acquisition card with the 16 bit resolution and 600 kSa/s per channel sampling. The whole experimental setup schematic is in Figure 3).

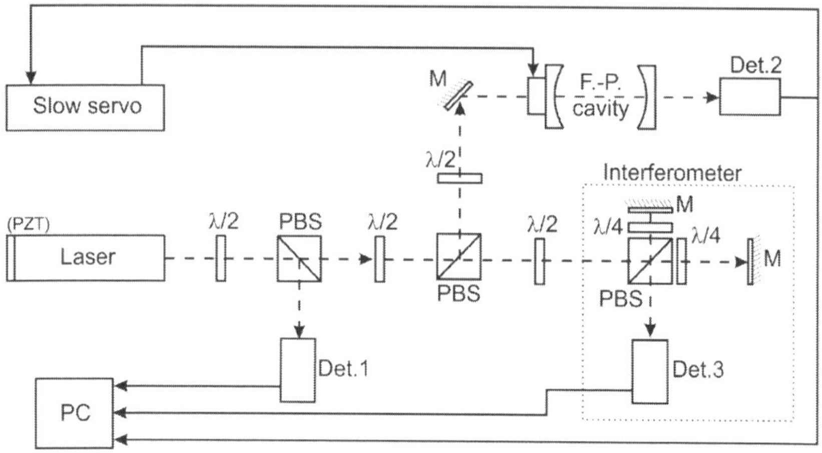

Figure 3. Schematic of the experimental setup for amplitude and frequency noise measurements. PBS-polarizing beam splitters, M-full reflective mirrors, Det-photodetectors (Det1 measuring amplitude noise, Det2 frequency noise throught the Fabry-Perot cavity, Det3 frequency noise throught the interferometer). PZT tuning option is included only in L1 laser and it was used only for evaluation of correct function of the frequency discriminators before the experiment itself.

The amplitude and frequency noise was measured for different free-running (unstabilized) laser sources (list of the tested lasers is at Table 3). However L5 commercial general purposes laser head contains thermal stabilization by the two-mode stabilization technique [18,19,51].

Table 3. A list of tested laser sources.

Laser no.	Description	Optical power [mW]
L1	Nd:YAG external cavity doubling, +noiseeater	20
L2	Nd:YAG external cavity doubling, −noiseeater	20
L3	Nd:YAG internal cavity doubling	30
L4	Nd:YAG, Simple DPSSL (diode pumped solid state laser)	50
L5	He-Ne, general purposes, thermally stabilized	5
L6	He-Ne, general purposes	0.8
L7	He-Ne, 25 years old general purposes	0.2
L8	He-Ne-I_2 laser standard	0.2

L1 (L2) laser is an ultra-stable ring Nd:YAG laser intended for metrology applications. It offers a "noise-eater" option which means switchable internal filter of amplitude noise. Measurement was done with the filter activated and also deactivated (highlighted as L1 and L2 respectively). L3 laser is an ultra-stable diode-pumped ring laser with an intracavity frequency doubling. Both L1 and L3 are primarily intended for saturated sub-Doppler spectroscopy in iodine vapor at 532 nm wavelength and are designed to operate as laser optical frequency standards [21,25–32]. L4 laser is a simple diode pumped solid state laser (DPSSL) with alignment-free monolithic resonator equipped with slow thermal frequency tuning option which allows linear absorption spectroscopy frequency stabilization technique. Some results of the long-term frequency stability of this laser stabilized by linear absorption technique are summarized in [5,36,38]. L5 and L6 lasers are commonly used commercial single-mode He-Ne lasers from different producers, L5 is equipped with thermal two-mode frequency stabilization, L6 is completely free-running. L7 is an approximately 25 years old commercial He-Ne laser, where degradation of the laser tube during long time period can be expected. Finally, L8 is He-Ne-I_2 laser—iodine stabilized optical frequency standard for 633 nm [16,17].

4. RESULTS AND DISCUSSION

Amplitude noise measurements (in range 0–100 kHz) were performed for free-running lasers directly by measurement of optical power fluctuations using the low noise photodetector after few hours of stable operation of all of the laser heads (Figure 4). The detection chain was not completely free of unwanted interfering electric signal with 20 kHz frequency coming from the switched laser power supplies. Another 50 Hz frequency component from the power supply network could not be completely avoided so these frequency components are present in measured spectra.

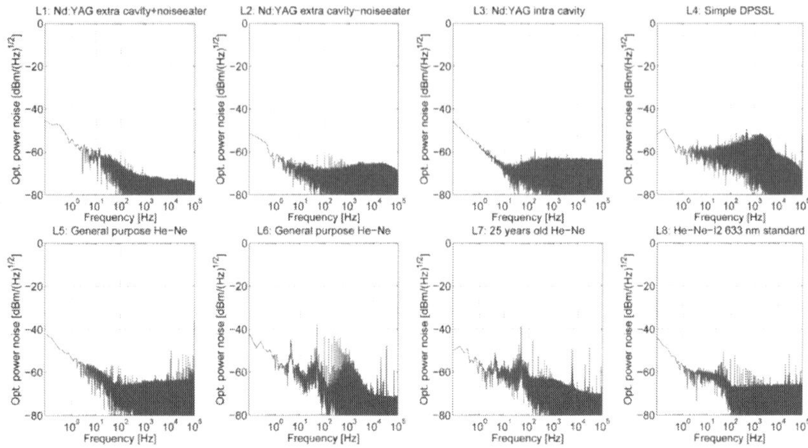

Figure 4. Optical power (amplitude) noise measurements of tested laser heads.

The best results were obtained with L1(2) and L3 (both of them designed for optical frequency standards) with noise floor level below -70 dBm/Hz$^{1/2}$ and -62 dBm/Hz$^{1/2}$, respectively, within the investigated bandwidth. L4 (simple DPSSL) shows low frequency fluctuations in the region of 0–10 kHz, caused most likely by acoustic noise interference due to mechanical vibrations of fan-cooled laser head. In case of He-Ne lasers the best result was obtained with L8 (He-Ne-I_2 standard), with noise floor at -65 dBm/Hz$^{1/2}$ and few harmonics (about -55 dBm/Hz$^{1/2}$), especially in the region of 35 kHz. Another of our He-Ne lasers shows amplitude fluctuations in range of -45 to -50 dBm/Hz$^{1/2}$. We assume that these fluctuations were mainly caused by electric interference from the switched power-supplies and in case of L7 also by ageing of old laser tube.

The frequency noise measurements (Figure 5) were done through the Fabry-Perot cavity frequency discriminator at the same sampling speeds as amplitude noise measurements (600 kSa/s) after the evaluation of the proper operation of the discriminator.

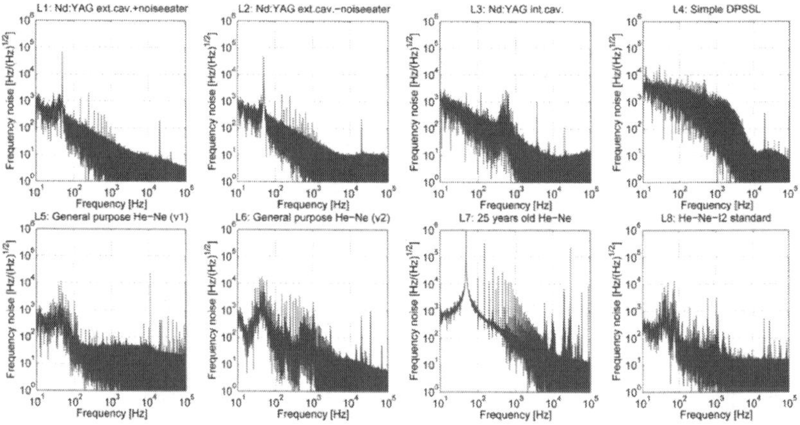

Figure 5. Frequency noise measurements of tested laser heads (Fabry-Perot cavity).

The RMS (root mean square) values of the frequency noise were calculated within different frequency bandwidths. Table 4 shows RMS values of frequency noise for spectral components above 10 Hz (to suppress an influence of control servo-loop of the Fabry-Perot cavity), Table 5 shows noise levels above 100 Hz to suppress the influence of 50 Hz spectral component present due to electric interference.

Table 4. RMS values of frequency noise *vs.* frequency bandwidth [kHz/(Hz)$^{1/2}$]. Starting frequency of the considered band is 10 Hz.

	Frequency noise bandwidth (from 10 Hz to ...)				
	100 Hz	1 kHz	10 kHz	100 kHz	300 kHz
L1	52.90	53.00	53.01	53.03	53.04
L2	50.70	50.83	50.85	50.91	50.94
L3	12.36	17.44	17.89	18.13	21.35
L4	73.42	95.56	96.80	96.86	96.87
L5	17.20	17.32	17.94	44.12	45.33
L6	33.88	35.55	36.53	39.32	39.39
L7	1,197.35	1,220.85	1,220.93	1,233.39	1,244.09
L8	24.37	24.71	25.13	25.35	26.00

Table 5. RMS values of frequency noise *vs.* frequency bandwidth [kHz/(Hz)$^{1/2}$]. Starting frequency of the considered band is 100 Hz.

	Frequency noise bandwidth (from 100 Hz to ...)			
	1 kHz	10 kHz	100 kHz	300 kHz
L1	3.14	3.37	3.60	3.72
L2	3.67	3.90	4.68	4.98
L3	12.31	12.92	13.26	17.42
L4	61.17	63.09	63.18	63.19
L5	2.8	5.11	40.63	41.94
L6	10.75	13.64	19.94	20.9
L7	238.38	238.79	296.00	337.81
L8	4.10	5.79	6.99	9.059

Considering the values above and the desired frequency bandwidth of the system, it is possible to evaluate the resulting noise of the measured displacement induced by the laser noise by the modifiedEquation (2). When Δf is the RMS value of the frequency noise within the considered bandwidth, the corresponding RMS value of the length variations at the output of the interferometer ("length noise") will be:

$$\Delta L = L_{DIFF} \times \left(\frac{f_0}{f_0 - \Delta f} - 1 \right) \tag{4}$$

where L_{DIFF} is length difference between reference and measuring paths of the interferometer and f_0 is the central optical frequency of the laser source. Some selected noise values for different paths differences and for different levels of frequency noise were computed and are shown in the Table 6.

When we put together calculated values from Table 6 and measured results from Tables 4 and 5 we can estimate a contribution of the laser frequency noise to the overall uncertainty of the interferometric distance measurement. For considered frequency bandwidth from 100 Hz–100 kHz and interferometer paths length difference 0.01 m we get RMS noise of the measured length of 8.30e-14 m for the L1 laser (4.68 kHz/(Hz)$^{1/2}$) to 5.25e-12 m length noise for the L7 laser (296 kHz/(Hz)$^{1/2}$). For the considered system frequency bandwidth from 100 Hz–300 kHz and interferometer paths length difference 1 m we get RMS noise of the measured length of 8.83e-12 m for the L1 laser (4.98 kHz/(Hz)$^{1/2}$) to 5.99e-10 m length noise for the L7 laser (337.81 kHz/(Hz)$^{1/2}$). Consequently, great care should be taken to frequency noise contribution to the measurement uncertainty.

Table 6. Exemplary computed RMS length noise ([m]) dependency on interferometer paths differences and the frequency noise of the laser.

	Δf [kHz]			
L_{DIFF} [m]	1e1	1e2	1e3	1e4
0.01	1.77e-13	1.77e-12	1.77e-11	1.77e-10
0.1	1.77e-12	1.77e-12	1.77e-10	1.77e-9
1	1.77e-11	1.77e-12	1.77e-9	1.77e-8

5. CONCLUSIONS

Our contribution deals with the interferometric measurements of lengths and discussion about aspects which influence the measurement results. We mainly concentrated on the short-term frequency noise of the laser head, which is one of the key parameters of the interferometer. In the experimental section the amplitude and frequency noise of different laser sources intended for interferometry were measured. The experimental setup with frequency discriminator represented by Fabry-Perot cavity with length stabilized through the slow servo-loop to the frequency of the investigated laser optical frequency was assembled. Measurements of amplitude and frequency noises were done simultaneously with using the high-speed digitization card. The correct operation of frequency discriminator was tested with the help of additional modulation of the laser frequency and the second part of the optical setup—the Michelson interferometer. The frequency measurements results show very good correlation between Fabry-Perot measurements and interferometer measurements. The best results of the amplitude noise perform at 532 nm Nd:YAG laser standards (L1,2 and L3), in case of He-Ne and 633 nm wavelength the best laser is L8 (He-Ne-I_2 standard). All of these lasers are intended to operate as laser standards for realization of fundamental etalons of length at 532 and 633 nm respectively in the laboratory environment. Their design is more precise, these lasers are more expensive in comparison to other tested laser sources. L4 (a simple DPSSL) contains internal servo loops for driving the laser and also a fan for cooling of the laser which caused huge frequency variations in the low frequency range. The worst frequency stability results corresponded to the L7 laser (an old one), mainly due to degradation due to the long history of this laser, old construction of the laser itself and its power supply. We computed achievable RMS values of the "length noise" for different levels of the frequency noise of the laser together with several measuring differences and reference paths of the interferometer. These results give us important information about the influence of the laser head's frequency noise properties to the interferometric measurement and the resolution and uncertainty contribution to the system. Exceptional interest should be taken on them especially for the setups with large interferometer paths difference. The next aspects we investigated are the resolution limit and bandwidth of the interferometer and its detection chain. We expect that greatest contributors to the frequency noise in the recordings are the switching power supplies of the laser heads. The results show better performance in frequency stability for measuring systems with Nd:YAG lasers in comparison to traditionally used He-Ne ones, especially for higher frequencies. This corresponds to measurements of long-term frequency stabilities which were done before [5,26–28]. Higher power available with Nd:YAGs lasers is not only high enough for feeding all of the interferometers needed especially in case of multidimensional systems but high power level incident on the photodetectors needs lower gain of the following amplifiers and thus improves the noise performance of the detection chain.

ACKNOWLEDGMENTS

The authors wish to express thanks for support to the grant projects from the Grant Agency of CR, projects: GPP102/11/P820 and GA102/09/1276, Academy of Sciences of CR, project: KAN311610701, Ministry of Education, Youth and Sports of CR, projects: CZ.1.05/2.1.00/01.0017 and LC06007, European Social Fund and National Budget of the Czech Republic, project: CZ.1.07/2.4.00/31.0016 and Technology Agency of CR, project: TA02010711.

REFERENCES

1. Korpelainen, V.; Seppa, J.; Lassila, A. Design and characterization of MIKES metrological atomic force microscope. Precis. Eng. 2010, 34, 735–744.

2. Jäger, G.; Gruenwald, R.; Manske, E.; Hausotte, T.; Fuessl, R. A nanopositioning and nanomeasuring machine: Operation-measured results. Nanotechnol. Precis. Eng. 2004, 2, 81–84.

3. Lazar, J.; Hrabina, J.; Sery, M.; Klapetek, P.; Cip, O. Multiaxis interferometric displacement measurement for local probe microscopy. Cent. Eur. J. Phys. 2012, 10, 225–231.

4. Kim, J.A.; Kim, J.W.; Park, B.C.; Eom, T.B. Measurement of microscope calibration standards in nanometrology using a metrological atomic force microscope. Meas. Sci. Technol. 2006, 17, 1792–1800.

5. Hrabina, J.; Lazar, J.; Klapetek, P.; Cip, O. Multidimensional interferometric tool for the local probe microscopy nanometrology. Meas. Sci. Technol. 2011, 22.

6. Wu, C.M.; Su, C.S.; Peng, G.S. Correction of nonlinearity in one-frequency optical interferometry.Meas. Sci. Technol. 1996, 7, 520–524.

7. Yacoot, A.; Downs, M.J. The use of x-ray interferometry to investigate the linearity of the NPL Differential Plane Mirror Optical Interferometer. Meas. Sci. Technol. 2000, 11, 1126–1130.

8. Cip, O.; Petru, F. A scale-linearization method for precise laser interferometry. Meas. Sci. Technol. 2000,11, 133–141.

9. Petru, F.; Cip, O. Problems regarding linearity of data of a laser interferometer with a single-frequency laser. Precis. Eng. 1999, 23, 39–50.

10. Eom, T.; Kim, J.; Jeong, K. The dynamic compensation of nonlinearity in a homodyne laser interferometer. Meas. Sci. Technol. 2001, 12, 1734–1738.

11. Rerucha, S.; Buchta, Z.; Sarbort, M.; Lazar, J.; Cip, O. Detection of interference phase by digital computation of quadrature signals in homodyne laser interferometry. Sensors 2012, 12, 14095–14112.

12. Ottonelli, S.; Dabbicco, M.; de Lucia, F.; di Vietro, M.; Scamarcio, G. Laser-self-mixing interferometry for mechatronics applications. Sensors 2009, 9, 3527–3548.

13. Olyaee, S.; Nejad, S.M. Nonlinearity and frequency-path modelling of three-longitudinal-mode nanometric displacement measurement system. IET Optoelectron 2007, 1, 211–220.

14. Badami, V.G.; Patterson, S.R. A frequency domain method for the measurement of nonlinearity in heterodyne interferometry. Precis. Eng. 2000, 24, 41–49.

15. Quinn, T.J. Mise-En-Pratique of the definition of the meter (1992). Metrologia 1994, 30, 523–541.

16. Navratil, V.; Fodrekova, A.; Gata, R.; Blabla, J.; Balling, P.; Ziegler, M.; Zeleny, V.; Petru, F.; Lazar, J.; Vesela, Z.; et al. International comparisons of He-Ne lasers stabilized with I-127(2) at lambda approximate to 633nm (July 1993 to September 1995)—Part III: Second comparison of Eastern European lasers at lambda approximate to 633 nm. Metrologia 1998, 35, 799–806.

17. Hrabina, J.; Petru, F.; Jedlicka, P.; Cip, O.; Lazar, J. Purity of iodine cells and optical frequency shift of iodine-stabilized He-Ne lasers. Optoelectron. Adv. Mat. 2007, 1, 202–206.

18. Ciddor, P.E.; Duffy, R.M. 2-Mode frequency stabilized He-Ne (633 Nm) lasers—studies of short-term and long-term Stability. J. Phys. E Sci. Instrum. 1983, 16, 1223–1227.

19. Ciddor, P.E.; Bruce, C.F. Long-Term stability of a thermally-stabilized He-Ne-Laser. Metrologia 1981,17, 17–18.

20. Simmons, J.D.; Hougen, J.T. Atlas of I2 spectrum from 19 000 to 18 000 Cm-1. J. Res. Natl. Bur. Stand. A Phys. Chem. 1977, 81, 25–80.

21. Hrabina, J.; Jedlicka, P.; Lazar, J. Methods for measurement and verification of purity of iodine cells for laser frequency stabilization. Meas. Sci. Rev. 2008, 8, 118–121.

22. Edlen, B. The refractive index of air. Metrologia 1966, 2, 71–80.

23. Zhang, J.; Lu, Z.H.; Menegozzi, B.; Wang, L.J. Application of frequency combs in the measurement of the refractive index of air. Rev. Sci. Instrum. 2006, 77.

24. Lazar, J.; Hola, M.; Cip, O.; Cizek, M.; Hrabina, J.; Buchta, Z. Displacement interferometry with stabilization of wavelength in air. Opt. Express 2012, 20, 27830–27837.

25. Galzerano, G.; Bava, E.; Bisi, M.; Bertinetto, F.; Svelto, C. Frequency stabilization of frequency-doubled Nd:YAG lasers at 532 nm by frequency modulation spectroscopy technique. IEEE Trans. Instrum. Meas. 1999, 48, 540–543.

26. Nevsky, A.Y.; Holzwarth, R.; Reichert, J.; Udem, T.; Hansch, T.W.; von Zanthier, J.; Walther, H.; Schnatz, H.; Riehle, F.; Pokasov, P.V.; et al. Frequency comparison and absolute frequency measurement of I-2-stabilized lasers at 532 nm. Opt. Commun. 2001, 192, 263–272.

27. Rovera, G.D.; Ducos, F.; Zondy, J.J.; Acef, O.; Wallerand, J.P.; Knight, J.C.; Russell, P.S. Absolute frequency measurement of an I-2 stabilized Nd:YAG optical frequency standard. Meas. Sci. Technol.2002, 13, 918–922.

28. Lazar, J.; Hrabina, J.; Jedlicka, P.; Cip, O. Absolute frequency shifts of iodine cells for laser stabilization. Metrologia 2009, 46, 450–456.

29. Picard, S.; Robertsson, L.; Ma, L.S.; Nyholm, K.; Merimaa, M.; Ahola, T.E.; Balling, P.; Kren, P.; Wallerand, J.P. Comparison of I-127(2)-stabilized frequency-doubled Nd:YAG lasers at the Bureau International des Poids et Mesures. Appl. Opt. 2003, 42, 1019–1028.

30. Galzerano, G.; Bava, E.; Bisi, M.; Bertinetto, F.; Svelto, C. Frequency stabilization of frequency-doubled Nd:YAG lasers at 532 nm. Proceedings of 1998 Conference on Precision Electromagnetic Measurements Digest, Washington, DC, USA, 6–10 July 1998; pp. 193–194.

31. Nyholm, K.; Merimaa, M.; Ahola, T.; Lassila, A. Frequency stabilization of a diode-pumped Nd:Yag laser at 532 nm to iodine by using third-harmonic technique. IEEE Trans. Instrum. Meas. 2003, 52, 284–287.

32. Holzwarth, R.; Nevsky, A.Y.; Zimmermann, M.; Udem, T.; Hansch, T.W.; von Zanthier, J.; Walther, H.; Knight, J.C.; Wadsworth, W.J.; Russell, P.S.J.; et al. Absolute frequency measurement of iodine lines with a femtosecond optical synthesizer. Appl. Phys. B. 2001, 73, 269–271.

33. Dai, G.L.; Pohlenz, F.; Danzebrink, H.U.; Xu, M.; Hasche, K.; Wilkening, G. Metrological large range scanning probe microscope. Rev. Sci. Instrum. 2004, 75, 962–969.

34. Cao, H.J.; Zang, E.J.; Zhao, K.; Zhang, X.B.; Wu, Y.X.; Shen, N.C. Frequency stabilization of a Nd:YAG laser to Doppler-broadened lines of iodine near 532 nm. Proc. SPIE 1998, 3549.

35. Lazar, J.; Cip, O.; Cizek, M.; Hrabina, J.; Sery, M.; Klapetek, P. Interferometer Controlled Positioning for Nanometrology. Proceedings of the 2nd International Conference Nanocon, Olomouc, Czech Republic, 12 October–14 December 2010; pp. 287–291.

36. Hrabina, J.; Lazar, J.; Klapetek, P.; Cip, O. AFM nanometrology interferometric system with the compensation of angle errors. Proc. SPIE 2011, 8082.

37. Otsuka, J.; Ichikawa, S.; Masuda, T.; Suzuki, K. Development of a small ultraprecision positioning device with 5 nm resolution. Meas. Sci. Technol. 2005, 16, 2186–2192.

38. Hrabina, J.; Lazar, J.; Cip, O.; Cizek, M. Laser source for interferometry in nanotechnology. Soc. Photo Opt. Inst. 2010, 7746, 53.

39. Clivati, C.; Mura, A.; Calonico, D.; Levi, F.; Costanzo, G.A.; Calosso, C.E.; Godone, A. Planar-Waveguide External Cavity Laser Stabilization for an Optical Link With 10(-19) Frequency Stability.IEEE Trans. Ultrason. Ferroelectr. Freq. Control 2011, 58, 2582–2587.

40. Lance, A.L.; Seal, W.D.; Labaar, F. Phase noise measurement systems. ISA Trans. 1982, 21, 37–44.

41. Buchta, Z.; Mikel, B.; Lazar, J.; Cip, O. White-light fringe detection based on a novel light source and colour CCD camera. Meas. Sci. Technol. 2011, 22.

42. Buchta, Z.; Rerucha, S.; Mikel, B.; Cizek, M.; Lazar, J.; Cip, O. Novel Principle of Contactless Gauge Block Calibration. Sensors 2012, 12, 3350–3358.

43. Yokoyama, T.; Araki, T.; Yokoyama, S.; Suzuki, N. A subnanometre heterodyne interferometric system with improved phase sensitivity using a three-longitudinal-mode He-Ne laser. Meas. Sci. Technol.2001, 12, 157–162.

44. Lazar, J.; Cip, O.; Cizek, M.; Hrabina, J.; Buchta, Z. Standing wave interferometer with stabilization of wavelength on air. TM Tech. Mess. 2011, 78, 484–488.

45. Lazar, J.; Cip, O.; Cizek, M.; Hrabina, J.; Buchta, Z. Suppression of air refractive index variations in high-resolution interferometry. Sensors 2011, 11, 7644–7655.

46. Birch, K.P.; Downs, M.J. Correction to the updated edlen equation for the refractive-index of air.Metrologia 1994, 31, 315–316.

47. Lazar, J.; Hola, M.; Cip, O.; Cizek, M.; Hrabina, J.; Buchta, Z. Refractive index compensation in over-determined interferometric system. Sensors 2012, 12, 14084–14094.

48. Quoc, T.B.; Ishige, M.; Ohkubo, Y.; Aketagawa, M. Measurement of air-refractive-index fluctuation from laser frequency shift with uncertainty of order 10(-9). Meas. Sci. Technol. 2009, 20.

49. Smid, R.; Cip, O.; Lazar, J. Precise length etalon controlled by stabilized frequency comb. Meas. Sci. Rev. 2008, 8, 114–117.

50. Lazar, J.; Cip, O.; Cizek, M.; Hrabina, J.; Buchta, Z. Interferometry with direct compensation of fluctuations of refractive index of air. Proc. SPIE 2010, 7746.

51. Balhorn, R.; Lebowsky, F.; Kunzmann, H. Frequency stabilization of internal-mirror helium-neon lasers. Appl. Opt. 1972, 11, 742–744.

5

GRAPHENE AND GRAPHENE NANOMESH SPINTRONICS

Junji Haruyama

Faculty of Science and Engineering, Aoyama Gakuin University, 5-10-1 Fuchinobe, Sagamihara, Kanagawa 252-5258, Japan

ABSTRACT

Spintronics, which manipulate spins but not electron charge, are highly valued as energy and thermal dissipationless systems. A variety of materials are challenging the realization of spintronic devices. Among those, graphene, a carbon mono-atomic layer, is very promising for efficient spin manipulation and the creation of a full spectrum of beyond-CMOS spin-based nano-devices. In the present article, the recent advancements in graphene spintronics are reviewed, introducing the observation of spin coherence and the spin Hall effect. Some research has reported the strong spin coherence of graphene. Avoiding undesirable influences from the substrate are crucial. Magnetism and spintronics arising from graphene edges are reviewed based on my previous results. In spite of carbon-based material with only sp^2 bonds, the zigzag-type atomic structure of graphene edges theoretically produces spontaneous spin polarization of electrons due to mutual Coulomb interaction of extremely high electron density of states (edge states) localizing at the flat energy band. We fabricate honeycomb-like arrays of low-defect hexagonal nanopores (graphene nanomeshes; GNMs) on graphenes, which produce a large amount of zigzag pore edges, by using a nonlithographic method (nanoporous alumina templates) and critical temperature annealing under high vacuum and hydrogen atmosphere. We observe large-magnitude ferromagnetism, which arises from polarized spins localizing at the hydrogen-terminated zigzag-nanopore edges of the GNMs, even at room temperature. Moreover, spin pumping effects are found for magnetic fields applied in parallel with the few-layer GNM planes. Strong spin coherence and spontaneously polarized edge spins of graphene can be expected to lead to novel spintronics with invisible, flexible, and ultra-light (wearable) features.

Keywords: spintronics; graphene; edges; spin polarization; ferromagnetism; magnetoresistance; rare-metal free

1. INTRODUCTION

Spintronics are highly promising as a key technology for next generation [1,2,3,4,5,6,7]. They have the following two prospects: (1) Zero-emission energy and (2) replacement of CMOS technology (*i.e.*, beyond CMOS). From the first prospect, electron spin currents carry and emit no energy and no heat. This strong advantage resolves heat problems in large-scale integration circuits, personal computers, and also any systems loading them. In particular, heat problems become much more significant in highly closed spaces (e.g., on aerospace and air planes). Spintronic devices and circuits must be extremely effective for such systems. From the second viewpoint, it is a desirable subject for human life to realize devices beyond CMOS FETs, which are approaching its integration and operation limits. Although many materials and technologies have challenged this, it has not yet been realized. Spintronic devices based on some kinds of ideas must realize this. For instance, operation utilizing spin flipping leads to extremely high switching devices (e.g., in pico-seconds), which overcomes operation speeds of COMS FETs and LSIs. Spin quantum bits also enable the treatment if large amounts of information by much smaller-scale integration compared to CMOS circuits. Therefore, spintronic devices are desired as future key technology.

Some kinds of structures have been developed for spintronics [1,2,3,4,5,6,7], such as giant magnetoresistance (GMR) [1], tunneling MR (TMR) [2,3], and spin-valve structures. In particular, TMR structure has realized high efficiency of TMR ratio. Even TMR ratio of as high as over 1000% has been obtained by using

CoFeB/MgO/CoFeB junction [4]. Spintronic devises have also based on a variety of materials, e.g., ferromagnetic metals (e.g., cobalt (Co), iron (Fe), chromium (Cr), manganese (Mn)) [1,2,3,4] and ferromagnetic semiconductors ((In, Mn), As) *etc.* [5,6,7]).

On the other hand, graphene, a carbon mono-atomic layer, has recently emerged followed by the discovery of an easy fabrication method, the so-called mechanical exfoliation of graphite. The field rapidly grew, and a Nobel Prize was awarded in 2010 for work related to this fabrication method. Here, strong spin coherence has been highly expected in graphene because of the weak spin-orbit interaction (SOI) and weak hyperfine interaction, which are unique to carbon atoms. Some works, however, reported on the weak spin coherence and the origins arising from substrate, impurities, and so on [8,9], while other works reported on the strong coherence, as predicted by theories [10,11]. Thus, the issue is still under debate. What is certain is that there exist undesirable influences originating from the substrate on the spin coherence, such as ripples, impurities, and defects. Because graphene is a mono-atomic layer directly fabricated on substrates, such factors significantly and certainly suppress the spin coherence. Avoiding these factors (e.g., by utilizing a hexagonal-boron-nitride (h-BN) substrate [12]) enables the fabrication of highly effective spintronic devices with a strong spin coherence.

As a recent exciting topic, phenomena associated with spin Hall Effect (SHE) [13] in graphenes have been recently reported. Realization of (quantum) SHE was theoretically predicted by resolving double degeneration of spin bands (e.g., by introducing spin orbit interaction (SOI)) and controlling two spins with opposite moments existing in different two bands only by applying electric fields [13,14,15,16]. Experimentally, observation of large spin diffusion current examined from a SHE-like method in high-quality graphenes on h-BN [12] and also pure SHE in hydrogenated graphenes with introducing SOI [17] has been recently confirmed. They are opening a door to novel all-carbon spintronics.

Moreover, in any of the spintronic devices mentioned above, rare magnetic elements are indispensable in order to provide polarized spins to the systems. In contrast, from a theoretical viewpoint, it is known that carbon-based sp^x-orbital systems can lead to the spontaneous emergence of electron spin polarization based on edge-localized electrons [18,19,20,21,22,23,24]. In particular, zigzag-atomic structure at graphene edges has been attracted nontrivial attention

[18,19,20,21,22,23,25,26,27,28,29,30,31,32,33,34,35,36,37,38,39,40,41,42,43]. The following two models predicted it.

One is the graphene nanoribbon (GNR) model, which assumes perfect edge atomic structures without any defects. It allows that the electron spins localizing at zigzag edges [18] to become stabilized toward polarization (*i.e.*, (anti)ferromagnetism) due to the exchange interaction between the two edges. This produces a maximum spin ordering in this orbital, in a GNR that is an one-dimensional strip line of graphene with edges on both longitudinal sides [18,19,20,21,22,23,24], in graphene nanomeshes (GNMs) with hexagonal nanopore arrays (Figure 1) [28,39] and in graphene nanoflakes [29]. This is similar to the case of Hund's rule for atoms. Moreover, spin configuration depends on kinds and number of foreign atoms (e.g., hydrogen (H) and oxygen (O)), which terminate edge carbon dangling bonds [20,41].

The other model is the so-called Lieb's theorem. Following it, the presence of low-concentration defects in ensemble of carbon atoms (e.g., graphene flakes) results in the appearance of net magnetism. It predicts the emergence of ferromagnetism by an increase in the difference between the number of removed A and B sites (Δ_{AB}) of the graphene bipartite lattice at zigzag edges [39,41]. The magnitude of ferromagnetism increases with increasing values of Δ_{AB}.

Although there are many reports for defect-based magnetism in graphene [44,45,46], very few experimental reports exist for purely zigzag-edge-derived magnetism. My group has successfully reported on observation of the edge polarized spins, ferromagnetism, and spin-based phenomena arising from zigzag pore edges by fabricating low-defect hydrogenated GNMs by a non-lithographic method [42,43]. Utilizing graphene edges spins leads to novel spintronic devices. For instance, the spin-filtering effect predicted that GNRs with antiferromagnetic spin alignment on two edges can only manipulate electron spins with the same moment and that the electron spins can be controlled by applying in-plane electric fields [18].

In the present manuscript, recent experimental advancement of spin-based phenomena in graphenes is reviewed in Section 2.1. Then, my group's previous experimental results about edge spins are reviewed inSection 2.2.

2. RECENT EXPERIMENTAL ADVANCEMENT OF SPIN-BASED PHENOMENA IN GRAPHENES

2.1. Advancement of Graphen Spintronics

2.1.1. Spin Coherence in Graphene

A large spin coherence length (>>1 μm) and time (>>1 ms) are expected for graphene, which allow for the realization of novel spintronic devices with a long lifetime and operation at room temperature. From a theoretical viewpoint, the following two causes are at least considered as the origin of the strong spin coherence in graphene: (1) the absence of the SOI due to the small mass of the carbon atom and (2) the weak hyperfine interaction due to the absence of the nuclear spins of the carbon nuclei (only 1% of the nuclei are ^{13}C and have spin). These causes are significantly different from those in other magnetic material systems (e.g., magnetic compound semiconductors and magnetic metals) consisting of heavy-mass atoms.

However, the spin lifetime (e.g., <1 ns) actually observed in some works is far from the predicted value, and the reasons for this observation have been discussed. For instance, ripples in the substrate, vacancies and adatoms may introduce non-uniform and local electric fields, resulting in the appearance of the SOI and the short spin lifetime. Charged impurities in the substrate and remote surface phonons may also weaken the spin coherence. Moreover, magnetic and paramagnetic defects may couple with the conducting spins and suppress the spin coherence. In order to clarify the origins, ref. [8] carried out the following well-known two quantum interference measurements, which are observed when the phase coherence length (l_Φ) is comparable to the device dimensions and disentangle the sources of magnetic and nonmagnetic decoherence: (1) universal conductance fluctuations (UCF) in which the dynamic conductance is quantized to a universal value independent of the sample size and geometrical scattering factors. Their statistics (e.g., fluctuation magnitude and dephasing rate) follow the theoretical predictions and are also determined by the magnetic dephasing mechanism; (2) Weak localization (WL) produces a maximum resistance and results in Anderson localization when the device size is much smaller than l_Φ. The WL fits to the average low-field magnetoconductance, which is determined only by nonmagnetic dephasing in graphene. The experiment combined UCF and WL measurements to separate and quantify the various sources of dephasing in graphene and revealed that both magnetic and nonmagnetic dephasing mechanisms limit the coherence in graphene below 1 K. The magnetic scattering rate was too large to be explained by remote magnetic moments, instead requiring that the magnetic defects are electronically coupled to graphene. Therefore, magnetic defects are the primary cause of spin relaxation, masking any potential effects of the SOI. Although the origins for magnetic defects in graphene were not clarified, it is quite important that the experiment revealed that magnetic defects are the dominant factor for spin decoherence.

The technique of pulsed electron spin resonance (ESR) spectroscopy is also useful [9,47] because it can avoid the influence of the electrode contacts on spin coherence. ESR spectroscopy can reveal the nature of edge spins and spin dynamics in detail through a broad multifrequency continuous wave. ESR experiments on pristine free-standing GNRs in [9] led to spin relaxation times on the order of microseconds and also discovered that the spin relaxation times are strongly temperature dependent, unlike the weak temperature dependence theoretically predicted. It was concluded that the spin relaxation times are determined by a tunneling level state-type relaxation process, typically observed for disordered materials in which the defect paramagnetic centers are not strongly bound to the main phonon bath, indicating that the phonon contribution is negligible in free-standing GNRs. On the basis of the data, it ruled out the undesirable contributions originating from the charged impurities in a substrate (causing strong impurity Coulomb scattering that limits the spin relaxation rate), as the GNRs are free-standing without being supported by an underlying substrate. This report also proved the above-mentioned causes arising from substrates for the suppression of spin coherence.

In contrast, some works reported on the spin coherence of graphene. For example, ref. [10] reported a spin coherence length as large as 2 μm and a spin coherence time as long as 150 ps at room temperature via non-local spin-valve measurements with Co electrodes on mono-layer graphene with an electron mobility of 2000 cm^2/Vs. These results were extracted from the Hanle-type spin precession measurements. The spin polarization of the ferromagnetic contacts was also calculated from the measurements to be around 10%. The report also implied extrinsic impurity scattering as an origin for spin scattering.

Moreover, reference [11] reported on anisotropic spin relaxation in graphene by similar measurements using a non-local spin valve structure. A comparison of the spin signals at $B = 0$ and $B = 2$ T showed a 20% decrease in the spin relaxation time for spins perpendicular to the graphene layer ($T_{\perp}$) compared to spins parallel to the layer ($T_{//}$). An analysis of the results was carried out in terms of the different strengths of the SO effective fields in the in-plane and out-of-plane directions. For example, if the SOI is of the Rashba or Dresselhaus type, the SO effective fields are exclusively in the graphene plane, and calculations showed that this should result in anisotropic spin relaxation in which $T_{\perp} = (1/2)T_{//}$. On the other hand, if the SO effective fields pointing out of the graphene plane dominate, it is expected that $T_{\perp} >> T_{//}$. This corresponded to the experimental result. The report also discussed the role of the Elliott-Yafet mechanism, in which spin scattering is induced by electron (momentum) scattering from impurities, boundaries, and phonons, and the Dyakonov-Perel mechanism, which results from the SO terms in the Hamiltonian of the clean material, for spin relaxation. This provided quite useful information for spin scattering for the SOI.

Consequently, spin coherence in graphene is still under debate. However, as I have mentioned above, it is crucial to avoid undesirable influences from the substrate on the spin coherence, which has been successfully realized by utilizing an h-BN substrate, which has very weak interaction with the graphene fabricated on it. In the next section, an example is considered in more detail.

2.1.2. Spin Hall Effect

The SHE is a phenomenon typically observed in a topological insulator [13,14,15,16] in which the electron spins are strongly preserved owing to topological reasons (e.g., the presence of the SOI), and the spin relaxation time is determined only by the uncertainty principle between energy and time. The SOI resolves the double degeneration of the energy bands and the opposite-moment spins (up and down moments) existing in the different two bands run toward opposite directions under applied electric fields, resulting in the appearance of the SHE. Thus, one can produce and control the Hall effect and spin currents without applying magnetic fields.

Two evident works associated with SHE in graphenes have been recently reported as mentioned above. One is the large spin diffusion current observed in SHE-like method in high-quality graphenes fabricated on h-BN substrate [12], while the other is observation of the intrinsic SHE in hydrogenated graphenes with introducing large SOI [17].

The former was observed in high-quality graphenes with electron mobility as high as ~150,000 cm^{-2}/Vs on h-BN substrate. It is well known that there is very weak interaction between h-BN and graphenes, resulting in less influence of substrate to the suppression of spin coherence mentioned above. It revealed true potential of graphenes similarly to the case of suspended graphenes. The experiments were carried out using electron spins with opposite moments, which are located in two different bands that were formed by Zeeman splitting. Making a constant current flow between two probes in four probe patterns (Hall pattern) under magnetic field over Zeeman energy, nonlocal resistance (R_{NL}) between other two probes was observed. Consequently, large peaks of R_{NL} were evidently confirmed, depending on temperature and magnetic fields. They could not be understood by conventional quantum Hall effect and could be rather similar to a SHE. The observed magnitude of the spin diffusion current was mostly 100 times greater than those in previous graphenes fabricated on conventional SiO_2 substrate. This strongly supports the relevance of the abovementioned origins for weak spin coherence caused by the interaction with substrate.

The result shows the true strong spin coherence of graphenes and also the high availability for spintronic devices. Quantum phenomena (e.g., fractional quantum Hall effect) in graphenes on h-BN substrate and also high electron motilities have been already reported, while reports of spin-based phenomenon are fewer in number. It reveals the essential natures of graphenes and enables an application to spintronic devices.

The latter observation of SHE was performed by terminating the individual carbon atoms of graphenes by hydrogen atoms. In the experiments, hydrogen silsesquioxane (HSQ) resist was used to precisely control a small amount of covalent H-termination of grahenes. After irradiation of the electron beam on the graphenes covered by HSQ, hydrogen atoms remained on the graphenes. As HSQ dose amount increases, the remaining amount of H-atoms increased linearly. After very weak hydrogenation ~0.02%, they observed a significant increase in R_{NL} (~400%), well above what can be accounted for by R_{Ohmic}. With increasing hydrogenation, the measured R_{NL} showed a steep increase, reaching 170 Ω at 0.05% hydrogenation. A strong increase of the R_{NL} was observed even at charge densities >1 × 10^{12} cm^{-2}. As the ohmic contribution to R_{NL} remained negligible over the entire hydrogenation rate, they argued that the only plausible explanation for the observed R_{NL} increase was the SHE. Moreover, the SHE was confirmed by the non-monotonic oscillatory behavior of the non-local signal in an applied in-plane magnetic field and also by the length, width and adatom density dependence of the non-local signal. From the length dependence of the non-local signal, they extracted a spin relaxation length of ~1 μm, a spin relaxation time of ~90 ps and a SOI strength of ~2.5 meV for samples with 0.05% hydrogenation. The SOI strength is actually significantly higher than ~10 μeV for conventional graphene with sp^2 orbitals.

The experiments, results, and analyses are quite interesting. The authors argued that an adatom locally breaks the reflection symmetry across the graphene plane, leading to an out-of-plane distortion by an angle ϕ relative to the plane (e.g., $\phi \approx 19.5°$ for a full sp^3) and mixing σ and π orbitals that are no longer orthogonal. Hence, the SO interaction becomes a first-order effect leading to a large enhancement of SO coupling for covalently bonded hydrogen impurities in graphene. However, the following questions still remain; (1) Why can carbon and hydrogen atoms with small mass lead to strong SOI? (2) How are the surface carbon atoms of bulk graphene covalently hydrogenated with the best angle for SP^3 orbitals? (3) How are the edge states and termination of edge dangling bonds correlated with the strong SOI?

From (1), even with formation of the sp^3 hybrid orbitals and out-of-plane distortion, the masses of hydrogen and carbon may be too small compared to heavy masses in conventional SOI cases. From (2), due to distortion and rippling from substrate, it seems to be hard to hydrogenate bulk graphenes with the best angle to form sp^3 orbitals. Why such C–H bonds were automatically formed remained in question. In this viewpoint, correlation of the edge dangling bonds is not clear for (3). Because the width of the graphenes used for the experiments are ~1 μm, they are not GNRs. Nevertheless, di-hydrogenation of the edge dangling bonds easily forms sp^3 orbitals compared to those in bulk. Thus, it is indispensable to reveal its contribution.

In any cases, these reports of the large spin diffusion currents of graphenes on h-BN substrate and the SHE realized by strong SOI in hydrogenated graphenes suggest high feasibility of graphene spins to novel spintronics. They must open a door to (magnetic-atom-free) carbon spintronics by manipulating electron spins by applying electric fields, after further investigation and development.

2.2. Magnetism and Spintronics on Pore Edge Spins in Graphene Nanomeshes

Many works have reported on magnetism arising from defects in graphene [44,45,46]. In contrast, very few works have experimentally reported on observation of magnetism and spin-based phenomena to truly arise from graphene zigzag edges. This is because edge-related phenomena are easily destroyed by disorder (damage, defects) and contamination introduced during the fabrication process (e.g., by lithographic methods). We have, therefore, developed two non-lithographic fabrication methods for low-defect graphene edges; (1) GNRs derived from unzipping of carbon nanotubes combined with air blow and three-step annealing [32] and (2) GNMs fabricated using nano-porous alumina template (NPAT) [42,43]. In the present section, observation of magnetism and spin-related phenomena are reviewed based on the latter one.

2.2.1. Ferromagnetism Arising from Zigzag-Type Pore Edges

2.2.1.1. Sample Fabrication for GNMs

GNMs, which have honeycomb-like array of hexagonal nanopores (Figure 1b,e), were fabricated on a large ensemble of mechanically exfoliated graphenes (or graphenes CVD-synthesized on SiC substrate) by using NPAT (Figure 1a) [48] as an etching mask (Figure 1c), following our previous nonlithographic method [27,36]. The NPAT is easily fabricated through self-organization by anodic oxidation of pure (99.99%) aluminum films. The graphene was carefully etched by optimized low-power Ar gas (e.g., 200–600 V for 10–40 min) so as to avoid giving damages (Figure 1d) and the naomesh of the NPAT was transferred to graphene. Then, the NPAT was detached from the surface of the fabricated GNM, either mechanically or chemically.

All the GNMs fabricated through these processes (including Figure 2c,f sample) were annealed at 800 °C in high vacuum (10^{-6} Torr) for 0.5–3 days with continual pumping of gas and, then, in hydrogen gas by the field-emission-type radical CVD system under pressure > 1 MPa at least for 3 h at for all the measurements. The first annealing is for deoxidization of the pore edges with recovering all damages and defects and is the key to forming zigzag pore edges by edge atomic reconstruction, while the second annealing is the key for termination of the carbon atoms at the pore edges by hydrogen atoms.

For observation of the features shown in Figure 2b,e, after the observation of the feature in Figure 2a, the sample was annealed at 800 °C in high vacuum (10^{-6} Torr) for 3 days with continual pumping of gas for de-hydrogenation of the edges and, then, in oxygen atmosphere for 1 h for oxidization of the pore edges. Magnetization measurements were carried out directly after the annealing.

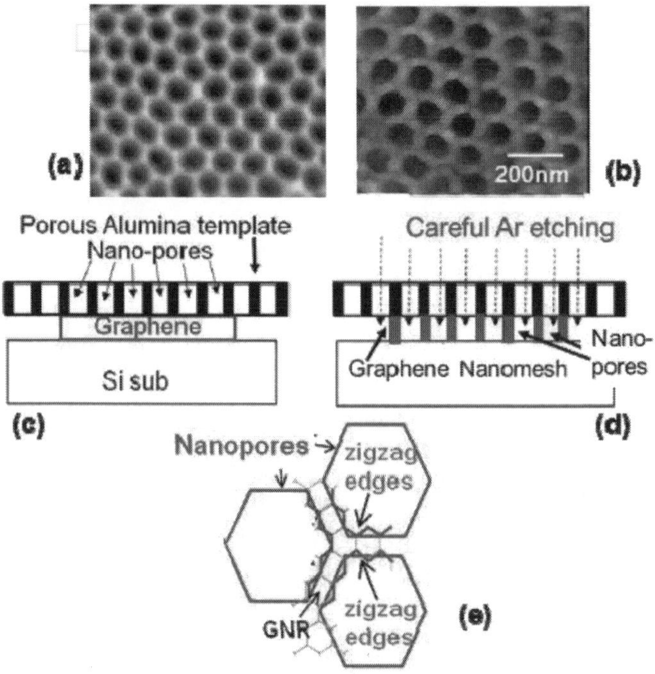

Figure 1. (**a**) SEM top view of a nano-porous alumina template (NPAT), which shows honeycomb-like array of hexagonal nano-pores; (**b**) AFM image of a graphene nanomesh (GNM) transferred following (**c**) and (**d**), which also proves the hexagonal shape of the pores (mean diameter $\phi \sim 80$ nm and mean inter-pore distance $W \sim 20$ nm); (**c,d**) Schematic cross sectional views of nonlithographic fabrication process of a GNM; (**c**) NPAT was placed on graphene as an etching mask; (**d**) The graphene is carefully etched by Ar gas, resulting in formation of a GNM; (**e**) Schematic view of a GNM with the zigzag-type pore-edge. Interpore regions can correspond to GNRs. Actual interpore regions include a larger number of hexagonal carbon unit cells per inter-pore region with length ~40 nm and $W \sim 20$ nm.

This GNM structure brings at least three large advantages as mentioned in the text.

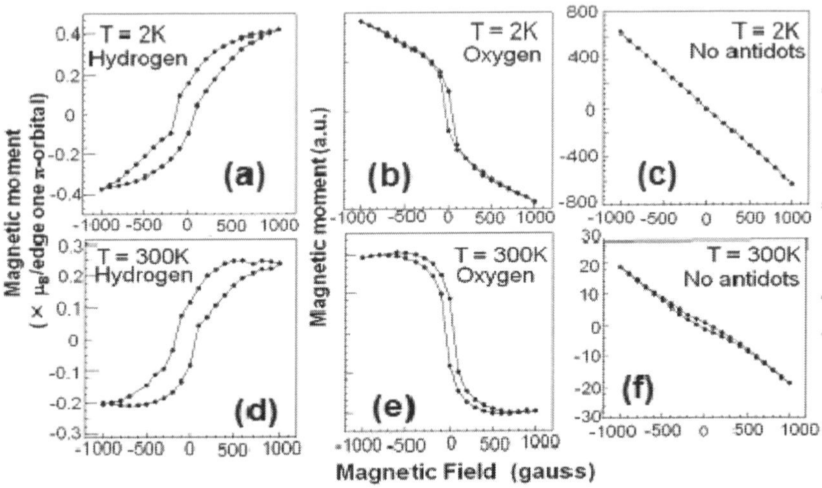

Figure 2. Magnetization of monolayer GNMs with ϕ ~ 80 nm and W ~ 20 nm for termination of different atoms. (**a**,**d**) hydrogen-terminated pore edges; (**b**,**e**) oxygen-terminated pore edges; and (**c**,**f**) bulk graphene without pore arrays. DC magnetization was measured by a superconducting quantum interference device (SQUID; Quantum Design). Magnetic fields were applied perpendicular to GNMs. The vertical axis in panels (**a**) and (**d**) denote magnetic moment per localized-edge π orbital, assuming mono-hydrogenation of individual edge carbon atoms as explained in text.

This method brings at least the following three significant benefits. (1) It gives fewer defects and contamination to the nanopore edges of GNMs because of the non-lithographic method; (2) The honeycomb-like array of hexagonal nanopores can result in the formation of a large ensemble of GNRs and pore edges with sufficient lengths (e.g., 40 nm in the present case), because one hexagonal nanopore can have six edges and the inter-pore regions can correspond to GNRs (Figure 1e). In the actual GNM, it is speculated that mixture of zigzag and armchair edges exists in one interpore GNR (one pore edge), as reported by previous STM observation [49]. Even so, a large number of GNRs in the present GNMs can yield a large area of assembled zigzag-edge GNRs. This is extremely effective in detecting small magnetic and electric signals arising from the pore edges; (3) When one would align the atomic structure of one pore edge to zigzag, the other five pore edges can automatically have zigzag atomic structure from a topological reason. At the current stage, this is impossible and we unintentionally form zigzag pore edges only via critical-temperature annealing, unlike in [35].

We have no direct evidence for the presence of zigzag-type pore edges. In [42], however, we indirectly proved presence of zigzag atomic structure at the pore edges by observation of the highly suppressed ratios of D/G peak heights (<0.2) in Raman spectroscopy, which were realized by reconstruction of edge atomic structure to zigzag type by the critical-temperature annealing. Indeed, at least two other groups reported similar correlation previously in intentionally fabricated zigzag-edged hexagonal nanopores [35] and also graphene flakes [40]. Conventionally, low D/G peak ratios can be evidence for low defects in samples. In contrast, they can also be evidence for zigzag pore edges when one dimensionality of the edges is taken into account [40].

As mentioned above, we could not intentionally form zigzag-type pore-edge atomic structure (e.g., using the advantage (3) mentioned above). In [30,31], however, it is suggested that the zigzag edge is the most stable chemically and that arm chair-based edges are reconstructed to zigzag after STM Joule heating for long edges of overlapped graphenes [30] and electron beam (EB) irradiation for pore edges [31]. This stability may be simply understood by the difference in the number of carbon atoms bonded to two neighboring carbon atoms (dangling bonds) for the zigzag edge (*i.e.*, one such atom) and arm chair edges (two such atoms) [31]. After removal of such atoms, the arm chair edge requires energy two-times larger than zigzag in order to repair the removed atoms and, thus, becomes unstable. In our system, we carried out high-temperature annealing for narrow ($W \sim 20$ nm) GNRs (*i.e.*, narrow inter-pore spacing). This should bring the energy similar to that in [30,31] and cause the reconstruction of pore edge atomic structures to zigzag.

Moreover, the observed anomalous periodic-magnetoresistance (MR) oscillation also showed the presence of localized electrons at the pore edges in *H*-terminated GNMs [42]. This also supports the presence of the *H*-terminated zigzag-type pore edges [42].

2.2.1.2. Magnetism on Pore Edge Termination by Different Foreign Atoms

The results of magnetization measured for different-foreign-atom terminated monolayer GNM are shown in Figure 2. A ferromagnetic-hysteresis loop with large amplitude is clearly observed only in *H*-terminated samples in Figure 2a,d, which showed the low D/G peak ratio values (<0.2). In addition to this sample, three other samples with low D/G peak heights in Raman spectroscopy exhibited similar ferromagnetism. In contrast, *O*-terminated GNMs exhibit a diamagnetism-like weak hysteresis loop (Figure 2b,e). This is consistent with [23], which reported that the formation of a spin-paired C–O chemical bond drastically reduces the local atomic magnetic moment of carbon at the zigzag edge of GNRs and suppresses the emergence of ferromagnetism.

Bulk graphenes without any pores and those assembled with NPATs show mostly no such features, even after H_2 annealing (Figure 2c,f), implying no contribution of parasitic factors (e.g., defects, impurities) of bulk graphenes. Moreover, the presence of fewer damages or impurities is reconfirmed in most the of bulk-graphene regions, because mechanically exfoliated bulk graphenes show an extremely low D/G peak heights (<<0.1) and a high 2D peak intensity in the Raman spectroscopy. These results strongly suggest that the observed ferromagnetism is associated with polarized spins localizing at the *H*-terminated zigzag-pore edges. It is surprising that the ferromagnetism observed at 2 K appears even at room temperature with a larger magnitude of the hysteresis loops (Figure 2d).

The correlation between the inter-pore spacing (corresponding to the width of the GNR, *W*; Figure 1e) and the magnetization is shown In Figure 3. It is found that the magnitude of hysteresis loop decreases with increasing *W*. In particular, it is revealed that the residual magnetization is inversely proportional to *W*value (inset of Figure 3c). This result is qualitatively consistent with the theories for the GNR model, according to which the edge spin stability and ordering of a zigzag-edge GNR are determined by the exchange interaction between the two edges, leading to vanishing of ferromagnetic edge spin ordering with increase of *W* [19,22]. Such a correlation cannot be understood by the ferromagnetism originating from the defects located only at pore edges or in the bulk graphene between pores. In the former case, ferromagnetism amplitude should be mostly independent of *W*, because defects are located only at pore edges. In the latter case, ferromagnetism amplitude should increase with an increase of *W*, because the density of the defects increases. Consequently, we conclude that the observed ferromagnetism is not due to parasitic origins (e.g., defects, impurities) but can be truly attributed to *H*-terminated zigzag pore edges.

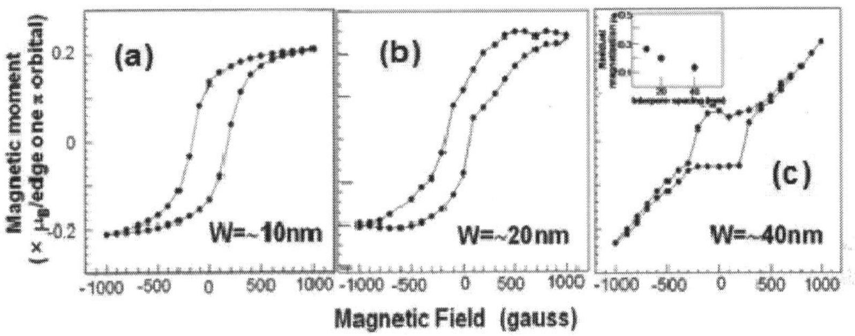

Figure 3. Correlation of the observed ferromagnetism as a function of the mean inter-pore spacing *W* at $T = 300$ K. Mean pore diameter ($\phi \sim$ 80 nm) was kept through all samples. For Figure 3a–c, difference in magnetic moment between upper and lower curves of hysteresis loop at$H = 0$ (residual magnetization $B_r \times 2$) is ~0.28 μ_B, ~0.2 μ_B, ~0.12 μ_B and the loop width at zero magnetic moment (coercivity $H_c \times 2$) is ~400 gauss, ~260 gauss, ~500 gauss, respectively. **Inset of** (**c**); B_r at 300 K as a function of *W*.

This is also consistent with MFM observation as shown in Figure 4. It clearly exhibits the presence of high density of polarized spins at the interpore GNR regions and also edge polarized spins at some pore edges. Because non-ferromagnetic samples show no such features, the observed ferromagnetism is attributed to these polarized spins. To date, approximately 50% of the samples, which include samples showing the low D/G peak heights, have shown ferromagnetism.

Disappearance of ferromagnetism in oxygen-terminated GNMs is interpreted as follows. Ref. [6] reported that the formation of a spin-paired C=O chemical bond drastically reduces the local atomic magnetic moment of carbon at the zigzag edge of GNRs and suppresses the emergence of ferromagnetism. The disappearance of ferromagnetism in Figure 2b,e is qualitatively consistent with this theory.

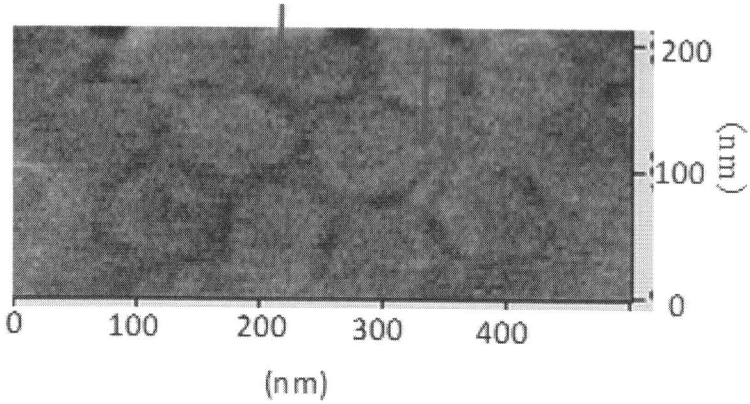

Figure 4. Magnetic force microscope (MFM) images of an H-terminated ferromagnetic GNM on SiC substrate. CoPtCr-coated Si probe was used for the measurements with a tapping mode. The interpore regions, which correspond to GNRs and exhibit darker color, imply high density of polarized spins. They directly suggest that the observed ferromagnetism originates from the GNRs. In particular, two arrows evidently imply the presence of edge polarized spins.

Moreover, in Figure 2b,e, even the diamagnetism of graphene mostly disappeared. One of the reasons for this is attributed to the formation of the nanomesh, because such an array drastically reduces the bulk graphene area available for the presence of loop currents to produce diamagnetism at the currently applied magnetic-field range (*i.e.*, corresponding to only GNRs with $W \sim 20$ nm between nanopores; Figure 1e). The radius of the cyclotron motion for electrons is given by $R_c = (\pi n_S)^{1/2}(h/2\pi)/eB$ [42]. By observing the magnetoresistance (*i.e.*, commensurability peak), we calculate n_S to be $\sim(4 \times 10^{11})$ cm^{-2} in the present GNMs. Based on this n_s value, R_c is calculated to be as large as ~400 nm, even for the currently applied largest magnetic field of 1000 gauss in Figure 2. Indeed, this R_c value is 20 times larger than $W \sim 20$ nm between the present nanopores, and it prohibits the emergence of loop currents for the formation of diamagnetism.

2.2.1.3. Theoretical Confirmation by Two Models: GNR Model and Lieb's Theorem

Here, two theoretical calculations quantitatively support our interpretation mentioned above. At first, employing the GNR model, which assumes pure zigzag pore edges (*i.e.*, without any defects) at all regions, enables the estimation of the magnetic moment of edge carbon atoms that contributes to the ferromagnetism observed in Figure 2. Assuming that only edge dangling bonds have localized spin moments, the magnetic moment per edge dangling bond prior to H termination is estimated to be $(1.2 \times 10^{-23})/(\mu_B = 9.3 \times 10^{-24})$ $\sim 1.3\ \mu_B$, where μ_B is the Bohr magneton, from the following.

The magnetic moment per edge dangling bond prior to H termination is estimated according to the following steps; (1) The total area of assembled bulk graphenes used for the pore array formation is ~ 4 cm^2; (2) The area of one hexagonal unit cell with a pore is $S = 6(3^{-1/2}/2)(a/2)^2 \sim 4300\ nm^2$, where a = [80 nm (pore diameter) + 20 nm (interpore spacing)]; (3) Thus, the total number of pores is $(4\ cm^2)/(4300\ nm^2) \sim 10^{11}$[(1)/(2)]; (4) The total number of dangling bonds per hexagonal pore is $(40\ nm)/(0.142\ nm \times 3^{1/2}) \times 6 = 166 \times 6 \sim 1000$; (5) The total number of edge dangling bonds of the GNM used for the SQUID measurement is 10^{14}[(3) × (4)].

Therefore, using (5), the saturation magnetization per edge dangling bond is estimated to be $1.2 \times 10^{-6}(emu) \times 10^{-3}/10^{14} = 1.2 \times 10^{-23}$ (J/T). Thus, the magnetic moment per edge dangling bond is estimated to be $(1.2 \times 10^{-23})/(\mu_B = 9.3 \times 10^{-24}) \sim 1.3\ \mu_B$, where μ_B is the Bohr magneton. Conventionally, GNRs with no H-termination of dangling bonds should show antiferromagnetism, theoretically. However, in the above estimation, the observed total magnetization was divided by estimated number of edge dangling bonds in the GNM, leading us to neglect this theory.

Next, after H annealing at critical temperature, edge dangling bonds of a GNR are terminated by H atoms [3,5,6,7,9]. The following three types of H terminations are theoretically possible. (1) All edge dangling bonds are each terminated by one H atom on the side of both zigzag edges. It provides a flat band for $2\pi/3 \leq k \leq \pi$ in the Brillouin zone. Electrons are well localized at the edges; (2) Each of the edge dangling bonds on one side are terminated by two H atoms (so that the edge carbon atom becomes tetrahedrally coordinated; a bearded edge), while the dangling bonds on the opposite side are terminated by a single H atom. The GNR provides a flat band for $0 \leq k \leq \pi$, resulting in a completely localized "on-bonding state" around the Fermi level (E_F). This leads to the spin polarization of all carbon atoms; (3) The double *H* atom termination of the zigzag-edge carbon atoms on both sides of a GNR provides a flat band for $0 \leq k \leq 2\pi/3$ and creates a modified zigzag edge.

The type of edge *H*-termination could not be observed in the present experiment. However, our case should correspond to case (1) from the following reason. The mono-*H* termination of the edge dangling bond decreases its magnetic moment to one μ_B. The magnetic moment of one localized-edge π orbital is, therefore, estimated to be as large as ($\sim$1.3 μ_B–1 μ_B) = $\sim$0.3 μ_B. This is in fairly good agreement with the theoretical contribution of the π-orbital state to the edge magnetic moment of $\sim$0.3 μ_B in a zigzag-edged GNR within the ferromagnetically ordered spin configuration [22].

As for the other model, we employ Lieb's model [39], which predicts the appearance of net magnetization arising from small defects in sublattice of carbon hexagonal cell (*i.e.*, missing A and B sites underlying bipartite lattice). Even a small defect may still remain in actual pore edges. In order to elucidate the influence of such residual small-volume edge disorder on magnetism of GNM, we performed systematic first-principles calculations of magnetic properties of quasi-GNR structures (Figure 5a, corresponding to the inter-pore region) based on Lieb's theorem [50], which introduces the slight disorder at the upper edge. Interestingly, the ground state of quasi-GNR structure turned out to be ferromagnetic in Figure 5b. The calculated net magnetic moment follows Lieb's theorem with local moments up to 0.2 μ_B/edge atom and depends on magnitude of the assumed edge disorder. These values agree fairly well with the value estimated from the GNR model.

Consequently, both models theoretically support that the observed ferromagnetism can originated from mono-hydrogenated zigzag nanopore edges. In order to determine which models are more relevant to the actual structures, observation of the pore edge atomic structures and number of the hydrogen atoms are indispensable. If the GNR model would be relevant, we have to assume presence of no defects at the pore edges. This seems to be realistic, because there are so many carbon atoms on one pore's edges and also so many nanopores in one GNM, as mentioned above. In contrast, if Lieb's theorem would be applicable, a small amount of defects should exist at the pore edges. However, following the above calculation, ferromagnetic spin alignment does not appear at the edge including defects in this case. This seems to be in poor agreement with the results of MFM observation mentioned above. Further experiments and comparison with theories are, therefore, indispensable.

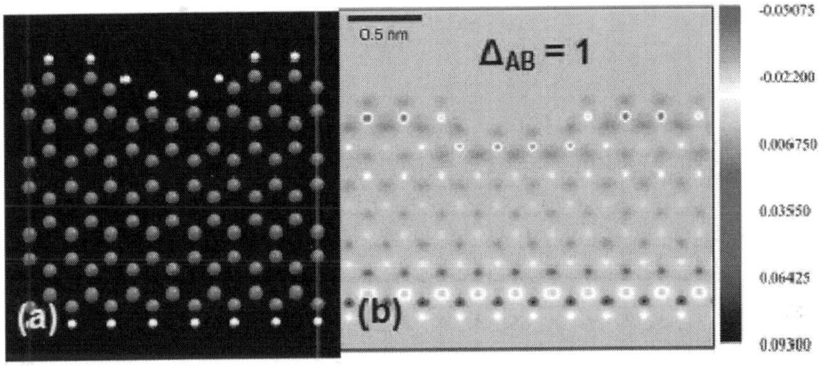

Figure 5. (**a**) Structure of mono-hydrogen-terminated interpore graphene nanoribbon (GNR) region, which introduces small curvature (defects) with Δ_{AB} = 1 *(i.e.*, the difference between the number of removed A and B sites of the graphene sublattices at zigzag edges) on the upper edge, used for first-principles calculations based on Lieb's theorem. The dark and white atoms are carbon and hydrogen, respectively; (**b**) Calculated spin-density distribution of quasi-GNR for (**a**), which gives the edge magnetic moment of 0.2 μ_B to the lower edge.

2.2.2. Spin-Based Phenomena in MR Measurements of Few-Layer GNMs

We have shown spontaneous spin polarization and ferromagnetism originating from the zigzag pore edges of monolayer GNMs. In contrast, the ferromagnetic GNMs demonstrated no clear electronic and MR features, because lithographical fabrication was used for formation of electrodes on the monolayer GNM. This method introduced contamination and disorder to the pore edges.

On the other hand, we previously confirmed periodic MR oscillations arising from electrons localizing at*H*-terminated pore edges in weak-ferromagnetic GNMs with ~10 layers, when magnetic fields were applied perpendicular to layers of the GNMs [42]. The observed oscillation period proved that the electron orbit localizing at the pore edges produced the oscillation with a period, which is shorter than those caused by free cyclotron-motion electrons due to applied fields. This result suggested that the contamination and disorder exist only around the most surface or the second-surface layers and the lower layers (e.g., the fifth layer) should mostly have no such influence. Thus, also in the present measurements, we measured MR behaviors of *H*-terminated GNMs with ~5 layers. Although magnitude of spin polarization at pore edges becomes weaker in the ~5 layer GNMs compared to that of monolayer GNMs, we confirmed that the GNMs actually exhibited evident ferromagnetism in magnetization measurement and proved presence of polarized edge spins [38].

Typical drain current (I_{sd}) *vs.* drain voltage (V_{sd}) and I_{sd} *vs.* back-gate voltage (V_{bg}) behaviors of a ~5 layer ferromagnetic GNM are shown in Figure 6a and Figure 6b, respectively. Figure 6a exhibits a strong zero-bias anomaly, which can be attributed to either single electron charging effect in narrow interpore GNR regions or high interface resistance between Au electrodes and GNM. Figure 6b shows an *n*-type semiconducting behavior also due to the narrow inter-pore GNR regions, which can introduce energy ban gaps [32]. Because the band gaps are attributed to confinement of electrons into the one-dimensional space of the inter-pore regions (GNRs) [32], it is not controversial to assume the presence of zigzag pore edges.

MR (R_{xx}) behaviors of two different-*W* ferromagnetic GNMs ($W \sim 30$ nm and ~40 nm) for inset of Figure 6a-pattern under a constant current (I_{sd}) mode of a four probe measurement are shown in Figure 6c,d, when magnetic fields are applied in parallel with the GNM planes ($B_{//}$). They show saw-tooth like MR oscillations, in which MR monotonically increases with increasing fields, while it abruptly drops at a critical field and starts to increase again following a repeated manner. In Figure 6c for GNM with $W \sim 30$ nm, at $B < -0.05$ and $0.15 < B$, the oscillation periods are larger than those in the two oscillations around $B = 0$. The periods for all oscillations are the same for increasing and decreasing *B*. In contrast, in Figure 6d for GNM with $W \sim 40$ nm, the oscillation becomes ambiguous. We call this saw-tooth like MR oscillation the spin pumping effect. The effect means a repeated cycle of gradual accumulation (increased number) of polarized spins and its abrupt emission (decreased number) at critical fields. No such anomalous behavior can be reported in any previous MR phenomena (e.g., ferromagnetic MR behavior, giant MR, tunnel MR, and spin valve effect), which is mentioned in the introduction.

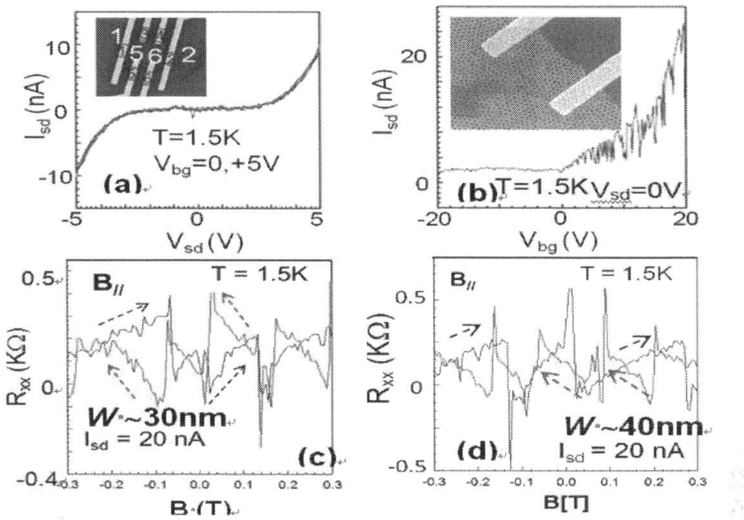

Figure 6. (**a**) Typical I_{sd} *vs.* V_{sd} and (**b**) I_{sd} *vs.* V_{bg} behaviors measured between electrodes 1 and 2 (inset of (**a**)) in a ferromagnetic GNM with ~5 layers. It exhibits an *n*-type semiconducting behavior due to the narrow inter-pore GNR regions (Figure 1e). **Inset of** (**a**): SEM image of Hall pattern (six probes) formed on a GNM. MR (R_{xx}) between electrodes 5 and 6 was measured under a constant I_{sd} of 20 nA between electrodes 1 and 2. **Inset of** (**b**): FESEM image of a GNM around electrodes, which shows presence of array of nanopores; (**c,d**) MR (R_{xx}) behaviors under parallel fields in two different W samples with ~5 layers (*i.e.*, (**c**) $W\sim$ 30 nm and (**d**) ~40 n, ϕ ~ 80 nm is common). V_{bg} was set to + 20 V. Arrows mean sequence of applied B (e.g., from −0.3 T to +0.3 T or from +0.3 T to −0.3 T).

In the present case, spin-polarized electrons localize at the pore edges under thermal equilibrium. However, under non thermal equilibrium caused by a constant current flow, the flat bands at the pore edges are modulated. Thus, some of the polarized electron-spins can no longer localize at the pore edges and flow through interpore bulk graphene regions. In our previous ~10-layer GNMs with $W\sim$ 20 nm [42], the spin pumping effect was not observed even under parallel fields, although MR oscillation arising from pore-edge localized electrons was found under perpendicular fields. This implies that enough inter-pore space W with bulk grapheme regions for electron transport is indispensable for the spin flow and detection of the spin pumping effect.

Qualitatively possible interpretation for the present spin pumping effect is explained as follows. When the applied magnetic field is increased, the number of the localized polarized-spins increases at the pore edges. Moreover, other free electrons, which are flowing through bulk graphene regions among the pores, also tend to accumulate and polarize, when they locate close to the flat energy band of the pore edges. However, the accumulation of polarized-spins saturate at the pore edges. Then, the excess spins are abruptly emitted at a critical field, as parallel magnetic fields increase further, because the flat band is modulated in the GNM plane by increasing parallel fields more or less. After the emission of the accumulated excess spins, the flat band partially recovers near to the initial condition and the pore edges can allow the accumulation of further spins. Then, the MR starts to increase again, resulting in a repeated saw-tooth like feature. In Figure 6d for the GNM with $W \sim 40$ nm, the saw-tooth like oscillation becomes ambiguous, as mentioned above. This is qualitatively consistent with a decrease in ferromagnetism amplitude shown in Figure 3c, that is, reduction of amplitude of polarized spins at the pore edges in monolayer GNMs. This suggests the presence of optimized W values (e.g., 30 nm), which is the most suitable for appearance of spin pumping effect, with respect to the amount of pore edge polarized spins and its flow through array of nanopores.

The large W values induce spin flow through the interpore GNR regions, while it weakens magnitude of the polarized spins at the pore edges. In contrast, the small W values induce the amount of the polarized spins at the pore edges. It also obstructs sufficient spin flow via the interpore regions between electrodes and makes the detection of the spin flow difficult. Further optimization of the structures and clarifying the exact mechanism, therefore, are expected in the near future.

3. CONCLUSIONS

Recent advancement of the spin-based research of graphenes and our experiments for the edge-based spin phenomena have been reviewed. The former promises precise manipulation of coherent spins by applying electric fields. Avoiding the negative influence of substrate-based factors is crucial (e.g., by using h-BN substrate). In the latter, we successfully fabricated low-defected mono-layer GNMs by using a non-lithographic method (*i.e.*, using NPATs) and found the spontaneous emergence of large-amplitude ferromagnetism, only when pore edges of the GNMs were *H*-terminated. Observed correlation between the inter-pore spacing and magnetism supported that it was attributed to electrons localizing at the zigzag pore-edges, which was in quantitative agreement with theories for pure GNR and defected-GNR models (Lieb's theorem). Moreover, the spin pumping effect was observed for fields applied in parallel with the GNM planes in few-layer GNMs with the optimized inter-pore spacing (~30 nm).

All-carbon spintronics is the key factor to realize magnetic rare-element free spintronic devices, which overcome the present material resource and environmental contamination problems. Strong spin coherence and its polarized edge spins of graphene must open the door to the novel spintronics with invisible, flexible, and ultra-light (wearable) features.

ACKNOWLEDGMENTS

The author thank H. Yang and M. Chshiev for the first principle calculation, K. Fujita, Y. Hashimoto, E. Endo, Y. Iye, S. Katsumoto, M. Yamamoto, S. Tarucha for the device fabrication processes, T. Ando, T. Enoki, M. Koshino, J. Akimitsu, T. Muranaka, H.Aoki, S. Roche, X. Jia, and M. S. Dresselhaus for their fruitful discussions and encouragement. This work at Aoyama Gakuin was partly supported by a Grant-in-aid for Scientific Research (Basic research A: 24241046) in MEXT, a High-Technology Research Center Project for private universities in MEXT, and also AFOSR grant.

CONFLICTS OF INTEREST

The authors declare no conflict of interest.

REFERENCES

1. Baibich, M.N.; Broto, J.M.; Fert, A.; Van Dau, F.N.; Petroff, F.; Etienne, P.; Creuzet, G.; Friederich, A.; Chazelas, J. Giant magnetoresistance of (001)Fe/(001)Cr magnetic superlattices. Phys. Rev. Lett. 1988,61, 2472.

2. Moodera, J.S.; Kinder, L.R.; Wong, T.M.; Meservey, R. Large magnetoresistance at room temperature in ferromagnetic thin film tunnel junctions. Phys. Rev. Lett. 1995, 74, 3273–3276.

3. Yuasa, S.; Nagahama, T.; Fukushima, A.; Suzuki, Y.; Ando, K. Giant room-temperature magnetoresistance in single-crystal Fe/MgO/Fe magnetic tunnel junctions. Nat. Mat. 2004, 3, 868–871.

4. Hayakawa, J.; Ikeda, S.; Lee, Y.M.; Matsukura, F.; Ohno, H. Effect of high annealing temperature on giant tunnel magnetoresistance ratio of magnetic tunnel junctions. Appl. Phys. Lett. 2006, 89, 232510.

5. Munekata, H.; Ohno, H.; von Molnar, S.; Segmüller, A.; Chang, L.L.; Esaki, L. Diluted magnetic III-V semiconductors. Phys. Rev. Lett. 1989, 63, 1849–1852.

6. Ohno, H.; Munekata, H.; Penney, T.; von Molnár, S.; Chang, L.L. Magnetotransport properties of p-type (In,Mn)As diluted magnetic III-V semiconductors. Phys. Rev. Lett. 1992, 68, 2664–2667.

7. Hai, P.N.; Ohya, S.; Tanaka, M.; Barnes, S.E.; Maekawa, S. Electromotive force and huge magnetoresistance in magnetic tunnel junctions. Nature 2009, 458, 489–492.

8. Lundeberg, M.B.; Yang, R.; Renard, J.; Folk, J.A. Defect-mediated spin relaxation and dephasing in graphene. Phys. Rev. Lett. 2013, 110, 156601.

9. Sun, Z.; Raji, A.O.; Zhu, Y.; Xiang, C.; Yan, Z.; Kittrell, C.; Samuel, E.L.G.; Tour, J.M. Large-area Bernal-stacked bi-, tri-, and tetralayer grapheme. Nano 2012, 6, 7615.

10. Tombros, N.; Jozsa, C.; Popinciuc, M.; Jonkman, H.T.; Van Wees, B.J. Electronic spin transport and spin precession in single graphene layers at room temperature. Nature 2007, 448, 571–574.

11. Tombros, N.; Tanabe, S.; Veligura, A.; Jozsa, C.; Popinciuc, M.; Jonkman, H.T.; Van Wees, B.J. Anisotropic spin relaxation in graphene. Phys. Rev. Lett. 2008, 101, 046601.

12. Abanin, D.A.; Morozov, S.V.; Ponomarenko, L.A.; Gorbachev, R.V.; Mayorov, A.S.; Katsnelson, M.I.; Watanabe, K.; Taniguchi, T.; Novoselov, K.S.; Levitov, L.S. Giant nonlocality near the Dirac point in graphene. Science 2011, 332, 328–330.

13. Murakami, S.; Nagaosa, N.; Zhang, S. Dissipationless quantum spin current at room temperature.Science 2003, 301, 1348–1351.

14. Kane, C.L.; Mele, E.J. Quantum spin hall effect in graphene. Phys. Rev. Lett. 2005, 95, 226801–226804.

15. Kane, C.L. Graphene and the quantum spin hall effect. J. Mod. Phys. B 2007, 21, 1155.

16. Schmidt, M.J.; Loss, D. Edge states and enhanced spin-orbit interaction at graphene/graphane interfaces. Phys. Rev. B 2010, 81, 165439.

17. Balakrishnan, J.; Koon, G.K.W.; Jaiswal, M.; Neto, A.H.C.; Özyilmaz, B. Colossal enhancement of spin-orbit coupling in weakly hydrogenated graphene. Nat. Phys. 2013, 9, 284–287.

18. Nakada, K.; Fujita, M.; Dresselhaus, G.; Dresselhaus, M.S. Edge state in graphene ribbons: Nanometer size effect and edge shape depend. Phys. Rev. B 1996, 54, 17954–17961.

19. Fujita, M.; Wakabayashi, K.; Nakada, K.; Kusakabe, K. Peculiar localized state at zigzag graphite edge.J. Phys. Soc. Jpn. 1996, 65, 1920–1923.

20. Kusakabe, K.; Maruyama, M. Magnetic nanographite. Phys. Rev. B 2003, 67, 092406.

21. Okada, S.; Oshiyama, A. Magnetic Ordering in Hexagonally Bonded Sheets with First-Row Elements.Phys. Rev. Lett. 2001, 87, 146803.

22. Lee, H.; Son, Y.; Park, N.; Han, S.; Yu, J. Magnetic ordering at the edges of graphitic fragments: Magnetic tail interactions between the edge-localized states. Phys. Rev. B 2005, 72, 174431.

23. Veiga, R.G.A.; Miwa, R.H.; Srivastava, G.P. Quenching of local magnetic moment in oxygen adsorbed graphene nanoribbons. J. Chem. Phys. 2008, 128, 201101.

24. Lee, H.; Park, N.; Son, Y.; Han, S.; Yu, J. Ferromagnetism at the edges of the stacked graphitic fragments: An ab initio study. Chem. Phys. Lett. 2004, 398, 207–211.

25. Enoki, T.; Takai, K. The edge state of nanographene and the magnetism of the edge-state spins. Sol. Stat. Comm. 2009, 149, 1144–1150.

26. Son, Y.; Cohen, M.L.; Louie, S.G. Energy gaps in graphene nanoribbons. Phys. Rev. Lett. 2006, 97, 216803.

27. Yang, L.; Park, C.; Son, Y.; Cohen, M.L.; Louie, S.G. Quasiparticle energies and band gaps in graphene nanoribbons. Phys. Rev. Lett. 2007, 99, 186801.

28. Shima, N.; Aoki, H. Electronic structure of super-honeycomb systems: A peculiar realization of semimetal/semiconductor classes and ferromagnetism. Phys. Rev. Lett. 1993, 71, 4389–4392.

29. Rosser, J.F.; Palacios, J.J. Magnetism in Graphene Nanoislands. Phys. Rev. Lett. 2007, 99, 177204.

30. Jia, X.; Hofmann, M.; Meunier, V.; Sumpter, B.G.; Campos-Delgado, J.; Romo-Herrera, J.M.; Son, H.; Hsieh, Y.; Reina, A.; Kong, J. Controlled formation of sharp zigzag and armchair edges in graphitic nanoribbons. Science 2009, 323, 1701–1705.

31. Girit, Ç.Ö.; Meyer, J.C.; Erni, R.; Rossell, M.D.; Kisielowski, C.; Yang, L.; Park, C.; Crommie, M.F.; Cohen, M.L.; Louie, S.G. Graphene at the edge: Stability and dynamics. Science 2009, 323, 1705–1708.

32. Shimizu, T.; Haruyama, J.; Marcano, D.C.; Kosinkin, D.V.; Tour, J.M.; Hirose, K.; Suenaga, K. Large intrinsic energy bandgaps in annealed nanotube-derived graphene nanoribbons. Nat. Nanotech. 2011,6, 45–50.

33. Han, M.Y.; Brant, J.C.; Kim, P. Electron transport in disordered graphene nanoribbons. Phys. Rev. Lett.2010, 104, 056801.

34. Wang, X.; Ouyang, Y.; Li, X.; Wang, H.; Guo, J.; Dai, H. Room-temperature all-semiconducting sub-10-nm graphene nanoribbon field-effect transistors. Phys. Rev. Lett. 2008, 100, 206803.

35. Krauss, B.; Nemes-Incze, P.; Skakalova, V.; Biro, L.P.; von Klitzing, K.; Smet, J.H. Raman scattering at pure graphene zigzag edges. Nano Lett. 2010, 10, 4544–4548.

36. Bai, J.; Zhong, X.; Jiang, S.; Huang, Y.; Duan, X. Graphene nanomesh. Nat. Nanotech. 2010, 5, 190–194.

37. Son, Y.W.; Cohen, M.L.; Louie, S.G. Half-metallic graphene nanoribbons. Nature 2006, 444, 347–349.

38. Otani, M.; Koshino, M.; Takagi, Y.; Okada, S. Intrinsic magnetic moment on (0001) surfaces of rhombohedral graphitee. Phys. Rev. B 2010, 81, 161403 (R).

39. Yang, H.; Chshiev, M.; Boukhvalov, D.W.; Waintal, X.; Roche, S. Inducing and optimizing magnetism in graphene nanomeshes. Phys. Rev. B 2011, 84, 214404.

40. You, Y.M.; Ni, Z.H.; Yu, T.; Shen, Z.X. Edge chirality determination of graphene by Raman spectroscopy. Appl. Phys. Lett. 2008, 93, 163112.

41. Soriano, D.; Leconte, N.; Ordejón, P.; Charlier, J.; Palacios, J.; Roche, S. Magnetoresistance and magnetic ordering fingerprints in hydrogenated graphene. Phys. Rev. Lett. 2011, 107, 016602.

42. Shimizu, T.; Nakamura, J.; Tada, K.; Yagi, Y.; Haruyama, J. Magnetoresistance oscillations arising from edge-localized electrons in low-defect graphene antidot-lattices. Appl. Phys. Lett. 2012, 100, 023104.

43. Tada, K.; Hashimoto, T.; Haruyama, J.; Yang, H.; Chshiev, M. Spontaneous spin polarization and spin pumping effect on edges of graphene antidot lattices. Phys. Status Solidi 2012, 249, 2491–2496.

44. Ning, G.; Xu, C.; Hao, L.; Kazakova, O.; Fan, Z.; Wang, H.; Wang, K.; Gao, J.; Qian, W.; Wei, F. Ferromagnetism in nanomesh graphene. Carbon 2013, 51, 390–396.

45. Nair, R.R.; Sepioni, M.; Tsai, I.; Lehtinen, O.; Keinonen, J.; Krasheninnikov, A.V.; Thomson, T.; Geim, A.K.; Grigorieva, I.V. Spin-half paramagnetism in graphene induced by point defects. Nat. Phys. 2012,8, 199–202.

46. Rao, S.S.; Jammalamadaka, S.N.; Stesmans, A.; Moshchalkov, V.V.; Van Tol, J.; Kosynkin, D.V.; Higginbotham-Duque, A.; Tour, J.M. Ferromagnetism in graphene nanoribbons: Split versus oxidative unzipped ribbons. Nano Lett. 2012, 12, 1210–1217.

47. Prével, B.; Benoit, J.-M.; Bardotti, L.; Mélinon, P.; Ouerghi, A.; Lucot, D.; Bourhis, E.; Gierak, J. Nanostructuring graphene on SiC by focused ion beam: Effect of the ion fluence. Appl. Phys. Lett. 2011,98, 83116.

48. Takesue, I.; Haruyama, J.; Kobayashi, N.; Chiashi, S.; Maruyama, S.; Sugai, T.; Shinohara, H. Superconductivity in entirely end-bonded multiwalled carbon nanotubes. Phys. Rev. Lett. 2006, 96, 057001.

49. Niimi, Y.; Matsui, T.; Kambara, H.; Tagami, K.; Tsukada, M.; Fukuyama, H. Scanning tunneling microscopy and spectroscopy of the electronic local density of states of graphite surfaces near monoatomic step edges. Phys. Rev. B 2006, 73, 085421.

50. Asano, H.; Muraki, S.; Endo, H.; Bandow, S.; Iijima, S. Strong magnetism observed in carbon nanoparticles produced by the laser vaporization of a carbon pellet in hydrogen-containing Ar balance gas. J. Phys. 2010, 22, 334209.

6

3D FINITE ELEMENT SIMULATION OF GRAPHENE NANO-ELECTRO-MECHANICAL SWITCHES

Jothiramalingam Kulothungan[1], Manoharan Muruganathan [1,*] and Hiroshi Mizuta [1,2]

[1]School of Materials Science, Japan Advanced Institute of Science and Technology, Nomi, Ishikawa 923-1292, Japan

[2]Nanoelectronics and Nanotechnologies Research Group, Faculty of Physical Sciences and Engineering, University of Southampton, Highfield, Southampton SO17 1BJ, UK

ABSTRACT

In this paper, we report the finite element method (FEM) simulation of double-clamped graphene nanoelectromechanical (NEM) switches. Pull-in and pull-out characteristics are analyzed for graphene NEM switches with different dimensions and these are consistent with the experimental results. This numerical model is used to study the scaling nature of the graphene NEM switches. We show the possibility of achieving a pull-in voltage as low as 2 V for a 1.5-μm-long and 3-nm-thick nanocrystalline graphene beam NEM switch. In order to study the mechanical reliability of the graphene NEM switches, von Mises stress analysis is carried out. This analysis shows that a thinner graphene beam results in a lower von Mises stress. Moreover, a strong electrostatic force at the beam edges leads to a mechanical deflection at the edges larger than that around the center of the beam, which is consistent with the von Mises stress analysis.

Keywords: graphene; nanoelectromechanical (NEM) switches; pull-in and pull-out characteristics; von Mises stress

1. INTRODUCTION

The standby power consumption of conventional complementary metal-oxide semiconductor (CMOS) circuits increases to the dynamic ON state level as they are scaled down to the scale of a few tens of nanometers [1]. On the other hand, nanoelectromechanical (NEM) switches are being investigated because of their promise for future low-power-consumption applications [2,3]. The switching operation of such devices is mainly based on electrostatic actuation, which leads to a very low leakage current and high ON/OFF ratios. They are also expected to achieve abrupt switching with subthreshold swing values less than 60 mV/dec [4]. Furthermore, NEM switches are expected to be robust against external disturbances such as radiation and temperature fluctuations, which makes them ideal for inhospitable environments [5]. On the other hand, the inherently low ON current in comparison with that of CMOS devices and the high pull-in voltage of conventional NEM switches restrict their use as a viable alternative for CMOS circuits and low-power applications [6,7].

Graphene is a two-dimensional material with excellent mechanical stability and electrical conductivity, and a high Young's modulus of ~1 TPa [8]. These outstanding properties of graphene make it a very promising material for high-performance NEM contact switches. Using graphene as a material for such devices can address some of the problems of conventional NEM switches by providing high reliability and a low pull-in voltage [9,10,11].

In this work, we present a three dimensional (3D) finite element method (FEM) simulation of double-clamped nanocrystalline graphene beam NEM switches. We focus on designing the graphene NEM switch in line with the experimental work [12]. The pull-in and pull-out characteristics from the simulation results are consistent with the experimental results. These models were used to study the double-clamped beam scaling on switching characteristics. After validating the simulation results for the pull-in voltage of the double-clamped graphene beam NEM switch with the experiment results, we studied the von Mises stress to evaluate the reliability of the switch. These results indicated that a longer and thinner graphene beam is more reliable. In addition, we also analyzed the effect of the applied electric field on the double-clamped graphene beam NEM switch.

2. DESCRIPTION OF THE DEVICE GEOMETRY

In this section, we briefly describe the geometry and operation principles of NEM switches. In our earlier work [12], we experimentally studied the switching operation of nanocrystalline graphene (NCG) beam NEM switches. The NCG was synthesized by direct deposition of NCG on an Si/SiO_2 substrate using plasma-enhanced chemical vapor deposition (PECVD) [13]. The NCG deposited by PECVD contains both sp^2- and sp^3-hybridized carbon atoms. The deposited NCG film is polycrystalline in nature [12]. The polycrystalline nanographene has randomly distributed grain orientation and size [14]. The mechanical behavior of NCG depends on both the grain misorientation and the grain boundary rotation [15]. Moreover, the mechanical strength of NCG depends on the arrangement of the defects in the NCG film [16]. The NCG sheets have almost constant fracture stress and strain, and the fracture strength is independent of the grain size [17]. The polycrystalline NCG sheets have a flaw-insensitive fracture mechanism [18]. Usually, fractures in NCG film originate from the grain boundary and propagate to the rest of the polycrystal [19]. The propagation of the fracture also depends on the grain orientation. Moreover, microcracks in the surfaces are initiated from topological defects in the NCG polycrystal, which coalesce to form a big crack, eventually leading to the breakdown of the NCG film. As NCG is a polycrystalline material, we considered it as an isotropic material in the FEM simulation of the nanocrystalline NEM switches. The device structure and dimensions are adopted from our previous experimental work [12]. We used a Young's modulus of 860 GPa for graphene in all the FEM simulations as reported by this experimental work. This value is comparable to the reported Young's modulus of 500 GPa for layered suspended graphene sheets of between 2 and 8 nm thickness [20]. For the FEM simulation of the experimental device, we considered a graphene beam of length L, width W, and thickness t, and a top metal electrode. The schematic representation of the device is shown in Figure 1. The double-clamped graphene beam NEM switch dimensions are detailed in Table 1. For each double-clamped graphene beam NEM switch, the air gap thickness changes owing to the natural buckling of the suspended graphene beam.

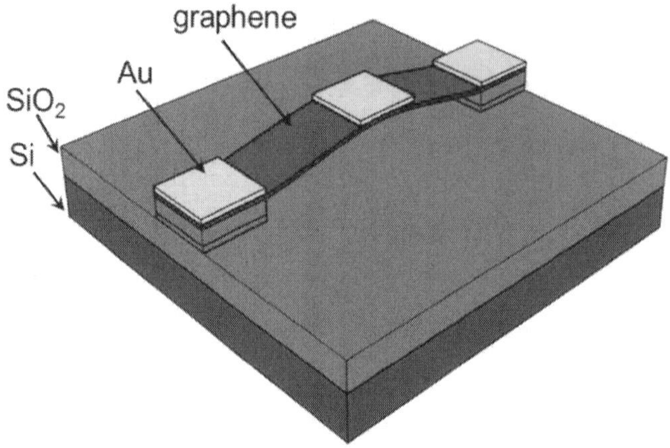

Figure 1. Schematic of double-clamped graphene beam NEM switch with top metal actuation electrode.

Table 1. NEM switch dimensions.

NEM Switch	Dimension			Air Gap Thickness (nm)
	Length (μm)	Width (μm)	Thickness (nm)	
A	1.5	0.5	9	75
B	1	0.5	9	50
C	0.8	0.5	9	45

The initial air gap of the device is g_0. As the voltage applied between the suspended graphene beam and the top electrode is increased, the resulting electrostatic force balances with the elastic restoring force of the deformed graphene. When the applied voltage reaches a critical point, the electrostatic force overwhelms the restoring force, which causes the graphene beam to be pulled towards the top electrode/fixed element, resulting in closing the switch and thereby leading to a sharp rise in the current flow through the device [21]. When the applied bias voltage is sufficiently reduced, the elastic restoring forces in the deformed active element pull the switch open, and thus the device operates as a volatile switch. A mechanical hysteresis is formed between the pull-in voltage and the pull-out voltage.

3. FINITE ELEMENT METHOD SIMULATION RESULTS AND DISCUSSION

3.1. Graphene NEM Switch Pull-In and Pull-Out Characteristics

The static electrical and mechanical characteristics of the NEM switch were simulated using the FEM-based CAD tool IntelliSuite (8.8.5.1, IntelliSense, Lynnfield, MA, USA) [22]. Figure 1 shows a schematic of our graphene beam NEM switch with a graphene beam connected to electrodes at each end. This graphene beam NEM switch features a metal top gate (actuation electrode), which enables the graphene beam to be pulled onto the gate when a voltage is applied, and then pulled away, disconnecting from the channel when voltage is no longer applied. In order to be consistent with our experimental device structure, we used the device dimensions mentioned in Table 1. Figure 2a shows the initial geometry of NEM switch A used in the FEM simulation. To analyze the pull-in and pull-out characteristics of the NEM switch, the voltage applied between the top electrode and the graphene beam was first increased until pull-in was confirmed and then decreased back to 0 V. Figure 2b illustrates the geometry of the pull-in state of the double-clamped graphene beam NEM switch. The color bar shows the displacement of the graphene beam with respect to the initial position.

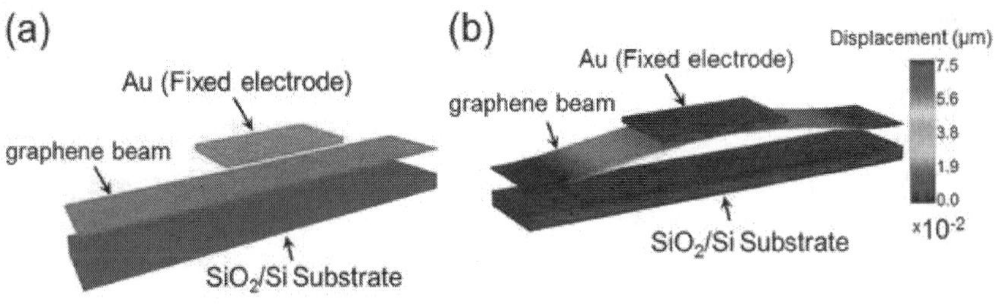

Figure 2. The geometry of NEM switch A. (**a**) Initial structure of double-clamped graphene beam NEM switch with a top metal electrode; (**b**) Pull-in state of the graphene beam; color bar indicates the relative displacement with respect to the initial condition.

Figure 3a shows the pull-in/pull-out characteristics obtained for the different graphene beam NEM switches mentioned in Table 1. The pull-in voltages for the graphene beam NEM switches A, B, and C are approximately 8.6, 13.2, and 20.8 V, respectively. The obtained pull-in voltage shown in Figure 3a is in good agreement with the experimentally reported pull-in voltages [13]. In order to clarify the impact of the thickness of the graphene beam, we conducted the FEM simulation for one of our experimental device structures (NEM switch A) with different graphene thicknesses of $t = 3$, 5, and 9 nm. Figure 3b shows the pull-in and pull-out characteristics for the graphene beam with different thicknesses for the voltage applied between the top gate and the graphene beam. When the thickness is reduced, the pull-in voltages are evaluated to be 8.6 V, 4.3 V, and 2.1 V for $t = 9$ nm, 5 nm, and 3 nm, respectively. This result shows a clear dependence of the pull-in voltage on the scaling of the thickness of the suspended graphene beam. Thus, we confirm that the introduction of thickness scaling leads to a reduction of the pull-in voltage.

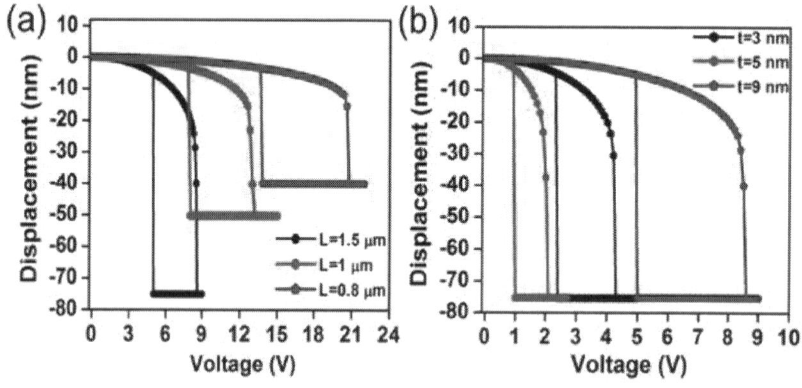

Figure 3. Pull-in and pull-out switching characteristics of the double-clamped graphene beam NEM switches. (**a**) Switching characteristics of graphene beam NEM switches A, B, and C; (**b**) Switching characteristics of graphene beam NEM switch A, with different graphene thicknesses of $t = 3$, 5, and 9 nm.

3.2. Von Mises Stress Analysis

The von Mises yield criterion is a general way to estimate the yield of any ductile material, such as metals [23,24]. The mechanical reliability of the graphene beam NEM switch can potentially be improved by properly choosing the switch dimensions. To quantitatively demonstrate the mechanical reliability of the double-clamped graphene beam NEM switch, we compared the maximum von Mises stress exerted along the length of the graphene beam [25,26]. The von Mises stress profile analysis is essential to comprehend the spatial variation of the stress generated on the suspended graphene owing to the applied voltage. A Cartesian coordinate system is used to represent the numerical coordinates on the suspended graphene beam. The stress profile was obtained after the pull-in state was achieved, giving the three-dimensional stress profile for the deformed graphene beam. However, the stress variation along the thickness is constant.

Figure 4 shows the von Mises stress for the different graphene beam NEM switches. The von Mises stress reaches the maximum value towards the ends of the graphene beam. When the length of the graphene beam is reduced, the von Mises stress is increased to the maximum value. As evident from Figure 4a, the device with the shortest graphene beam length has the maximum probability of failure. When the thickness of the graphene beam is scaled for the fixed length of the beam, the von Mises stress is reduced as the thickness is reduced. The results suggest that NEM switch A is at least as reliable as NEM switches B and C.

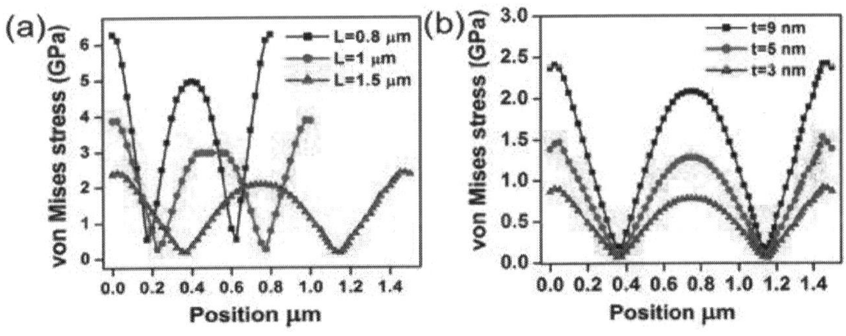

Figure 4. The von Mises stress of double-clamped graphene beam NEM switches. (**a**) von Mises stress of graphene NEM switches A, B, and C; (**b**) von Mises stress of graphene beam NEM switch A, with different graphene thicknesses of $t = 3$, 5, and 9 nm.

NEM switch C has a maximum von Mises stress of 6.2 GPa. When the length of the graphene beam is increased to 1 μm and 1.5 μm, it leads to a decrease in the von Mises stress of 3.8 GPa and 2.3 GPa, respectively. Furthermore, the maximum stress for NEM switch A is

2.5 GPa; when the thickness of the graphene is reduced to 5 nm and 3 nm, the stress is decreased to 1.4 GPa and 0.9 GPa, respectively. Figure 5illustrates the top view of the von Mises stress contour plot of the graphene beam. It is evident from the contour plot that the von Mises stress is highest nearer to both fixed ends of the beam. The von Mises stress reaches the minimum value between the fixed end of the beam and the center of the beam in the pull-in state. If we examine the von Mises stress across the beam carefully, then we can observe higher stresses at the edges of the graphene beam compared to those at the center of the beam. In order to clarify this point, 3D electric field distributions in the NEM switch were carried out in the pull-in state.

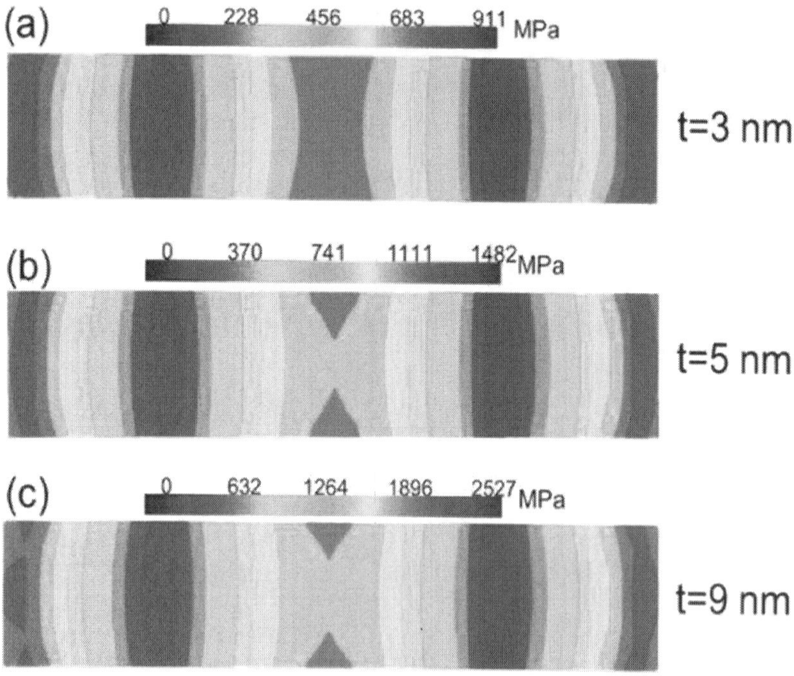

Figure 5. Contour plot (top view of graphene beam) of von Mises stress for the NEM switch A, with different graphene beam thicknesses of t = 3, 5, and 9 nm.

3.3. Three-Dimensional Electric Field Distribution and Its Role in Graphene Beam NEM Switch Operation

To analyze the impact of the applied electric field on the double-clamped graphene beam NEM switch, we made the same model in COMSOL Multiphysics (5.1, COMSOL Inc., Burlington, MA, USA) [27]. The NEM switch was built inside a vacuum environment. The model was meshed with triangular mesh elements to reduce the computational complexity. The density of the mesh was varied adaptively in order to study the structural displacement of the graphene beam. In this simulation, the actuation electrode was kept at the bottom and the graphene beam was placed at the top. For the electric field analysis at different voltages, a constant bias of 0 V was applied to the bottom electrode (Au) and the voltage applied at the top electrode (graphene beam) was swept. The potential, *V*, and the electric field, *E*, in the free space can be obtained by solving Poisson's equation [28]. Figure 6a shows the cross-sectional view of the electric field distribution across the center of the NEM switch for the applied

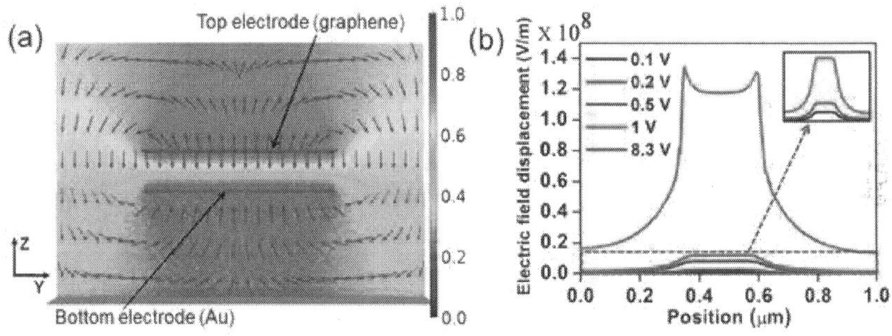

Figure 6. Two-dimensional electric field distribution across the center of the NEM switch. Dimensions of the graphene beam (top electrode) are equivalent to those of NEM switch A. (**a**) The electric field distribution across the switch at the center of the beam. Arrows indicate the electric field direction; (**b**) The electric field strength in the *Z* direction at 5 nm above the bottom electrode at different voltages applied between the bottom and top electrodes. Inset shows zoomed-in version of electric field distribution for lower actuation voltages as indicated by the dashed box; the scale is the same as (**b**).

voltage of 1 V to the bottom electrode. The dimensions of the graphene beam are equivalent to those of NEM switch A. Arrows in this plot show the electric field lines directions. At the center of the beam, the electric field lines are distributed vertically. The orientation of the electric field distribution is gradually changed to the horizontal direction towards the edges of the beam. At both edges of the beam, the electric field is distributed more horizontally in the outward direction from the center of the beam.

Figure 6b illustrates the one dimensional (1D) electric field strength in the *Z* direction at 5 nm above the bottom electrode for the different applied voltages. Consistent with Figure 6a, the electric field strength is highly concentrated at the edges of the graphene beam. These results demonstrate that the downward component of the electrostatic force acting on the edges of the graphene beam is higher than that at the center of the beam. In order to analyze this edge field termination effect, mechanical deflection analysis of NEM switch A with a 3 nm graphene thickness was done. Figure 7 shows the displacement of the graphene beam nearer to the pull-in state. If we consider the edge of the graphene beam, then the downward bend of the beam edges is apparent. This is also consistent with a higher von Mises stress at the beam edges.

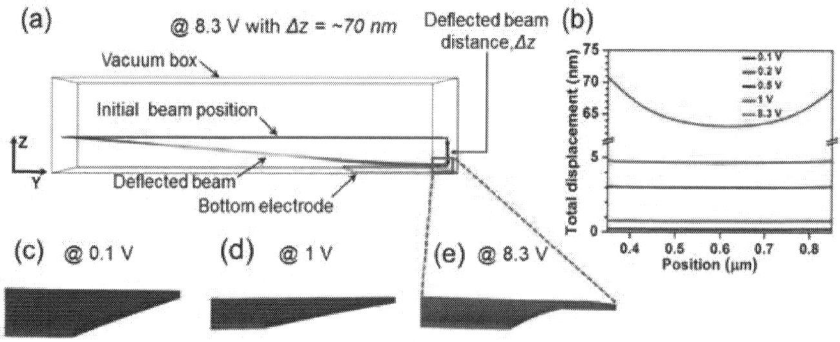

Figure 7. Effect of the beam edge electric field termination at 8.3 V. (**a**) The displacement of the beam at 8.3 V shown as side view; (**b**) The displacement of the beam from initial position at different applied voltages. Birds-eye cross-sectional view of the graphene beam at (**c**) 0.1 V; (**d**) 1 V; and (**e**) 8.3 V.

4. CONCLUSIONS

In this paper, we have studied the electro-mechanical switching and mechanical reliability of graphene beam NEM switches by FEM simulations. To evaluate the mechanical reliability of a graphene beam NEM switch, we scaled the length and thickness of the graphene beam and studied the von Mises stress for each structure. This analysis showed that the graphene beam NEM switch with a longer length of 1.5 μm and a thickness of 3 nm has a pull-in voltage of 2 V. The electrostatic force concentration at the edges of the graphene beam leads to more mechanical deflection at the edges than at the center of the beam.

ACKNOWLEDGMENTS

This work was supported by the Grant-in-Aid for Scientific Research No. 25220904 from the Japan Society for the Promotion of Science.

AUTHOR CONTRIBUTIONS

Manoharan Muruganathan conceived the concept; Jothiramalingam Kulothungan performed the simulations; Jothiramalingam Kulothungan, Manoharan Muruganathan, and Hiroshi Mizuta analyzed the data and wrote the paper; Hiroshi Mizuta supervised the project. All authors equally contributed to the results analysis and preparation of the manuscript.

CONFLICTS OF INTEREST

The authors declare no conflict of interest.

REFERENCES

1. Chui, C.O.; Shih, K.H.; Shoorideh, K. Low Dissipation Nanoscale Transistor Physics and Operations. In Proceedings of the 9th International Conference on Solid-State and Integrated-Circuit Technology (ICSICT 2008), Beijing, China, 22–23 October 2008; pp. 29–32.

2. Peschot, A.; Qian, C.; Liu, T.-J.K. Nanoelectromechanical switches for low-power digital computing.Micromachines 2015, 6, 1046–1065.

3. Dadgour, H.F.; Banerjee, K. Design and Analysis of Hybrid NEMS-CMOS Circuits for Ultra Low-Power Applications. In Proceedings of the 44th ACM/IEEE Design Automation Conference, San Diego, CA, USA, 4–8 June 2007; pp. 306–311.

4. Loh, O.Y.; Espinosa, H.D. Nanoelectromechanical contact switches. Nat. Nanotechnol. 2012, 7, 283–295.

5. Lee, T.-H.; Bhunia, S.; Mehregany, M. Electromechanical computing at 500 °C with silicon carbide.Science 2010, 329, 1316–1318.

6. Choi, S.-J.; Ahn, J.-H.; Han, J.-W.; Seol, M.-L.; Moon, D.-I.; Kim, S.; Choi, Y.-K. Transformable functional nanoscale building blocks with wafer-scale silicon nanowires. Nano Lett. 2011, 11, 854–859.

7. Loh, O.; Wei, X.; Sullivan, J.; Ocola, L.E.; Divan, R.; Espinosa, H.D. Carbon-Carbon contacts for robust nanoelectromechanical switches. Adv. Mater. 2012, 24, 2463–2468.

8. Geim, A.K.; Novoselov, K.S. The rise of graphene. Nat. Mater. 2007, 6, 183–191.

9. Li, P.; You, Z.; Haugstad, G.; Cui, T. Graphene fixed-end beam arrays based on mechanical exfoliation.Appl. Phys. Lett. 2011, 98, 253105.

10. Liu, X.; Suk, J.W.; Boddeti, N.G.; Cantley, L.; Wang, L.; Gray, J.M.; Hall, H.J.; Bright, V.M.; Rogers, C.T.; Dunn, M.L.; et al. Large arrays and properties of 3-terminal graphene nanoelectromechanical switches. Adv. Mater. 2014, 26, 1571–1576.

11. Sun, J.; Wang, W.; Muruganathan, M.; Mizuta, H. Low pull-in voltage graphene electromechanical switch fabricated with a polymer sacrificial spacer. Appl. Phys. Lett. 2014, 105, 33103.

12. Sun, J.; Schmidt, M.E.; Muruganathan, M.; Chong, H.M.H.; Mizuta, H. Large-scale nanoelectromechanical switches based on directly deposited nanocrystalline graphene on insulating substrates. Nanoscale 2016, 8, 6659–6665.

13. Schmidt, M.E.; Xu, C.; Cooke, M.; Mizuta, H.; Chong, H.M.H. Metal-free plasma-enhanced chemical vapor deposition of large area nanocrystalline graphene. Mater. Res. Express 2014, 1, 25031.

14. Wu, J.; Wei, Y. Grain misorientation and grain-boundary rotation dependent mechanical properties in polycrystalline graphene. J. Mech. Phys. Solids 2013, 61, 1421–1432.

15. Song, Z.; Artyukhov, V.I.; Yakobson, B.I.; Xu, Z. Pseudo hall–petch strength reduction in polycrystalline graphene. Nano Lett. 2013, 13, 1829–1833.

16. Wei, Y.; Wu, J.; Yin, H.; Shi, X.; Yang, R.; Dresselhaus, M. The nature of strength enhancement and weakening by pentagon–heptagon defects in graphene. Nat. Mater. 2012, 11, 759–763.

17. Yang, Z.; Huang, Y.; Ma, F.; Sun, Y.; Xu, K.; Chu, P.K. Size-dependent deformation behavior of nanocrystalline graphene sheets. Mater. Sci. Eng. B 2015, 198, 95–101.

18. Zhang, T.; Li, X.; Kadkhodaei, S.; Gao, H. Flaw insensitive fracture in nanocrystalline graphene. Nano Lett. 2012, 12, 4605–4610.

19. Shekhawat, A.; Ritchie, R.O. Toughness and strength of nanocrystalline graphene. Nat. Commun. 2016,7, 10546.

20. Frank, I.W.; Tanenbaum, D.M.; van der Zande, A.M.; McEuen, P.L. Mechanical properties of suspended graphene sheets. J. Vac. Sci. Technol. B 2007, 25, 2558–2561.

21. Zhang, W.-M.; Yan, H.; Peng, Z.-K.; Meng, G. Electrostatic pull-in instability in MEMS/NEMS: A review. Sens. Actuators Phys. 2014, 214, 187–218.

22. IntelliSense-Total MEMS Solutions-IntelliSuite Software. Available online: http://www.intellisense.com (accessed on 11 August 2016).

23. Beer, F. Mechanics of Materials, 2nd ed.; McGraw Hill: New York, NY, USA, 2002.

24. Young, W.C. Roak's Formulas for Stress and Strain, 7th ed.; McGraw Hill: New York, NY, USA, 2002.

25. Kim, D.H.; Kim, M.W.; Jeon, J.W.; Lim, K.S.; Yoon, J.B. Mechanical Reliability of a digital micromirror with interdigitated

cantilevers. J. Microelectromechanical Syst. 2010, 19, 1197–1206.

26. Boldeiu, G.; Vasilache, D.; Moagar, V.; Stefanescu, A.; Ciuprina, G. Study of the von Mises Stress in RF MEMS Switch Anchors. In Proceedings of the 2015 International Semiconductor Conference (CAS), Sinaia, Rome, 12–14 October 2015; pp. 219–222.

27. COMSOL Multiphysics® and COMSOL Server™. Available online: http://www.comsol.com (accessed on 11 August 2016).

28. Kambali, P.N.; Pandey, A.K. Capacitance and Force Computation Due to Direct and Fringing Effects in MEMS/NEMS Arrays. IEEE Sens. J. 2016, 16, 375–382.

7

MINIATURIZED WIDE-RANGE FIELD-EMISSION VACUUM GAUGE

Amra Avdić[1] , Anna Maria Lausch[1], Emmerich Bertagnolli[1] and Alois Lugstein[1]

[1] Vienna University of Technology, Institute of Solid State Electronics, Vienna, Austria

ABSTRACT

Miniaturized vacuum gauges (MVGs) for the measurement range 5.7times10^{-7} to 1.1times10^{-2} mbar were fabricated in a self-aligned approach using focused ion beam (FIB) nanoma-chining and reactive ion etching (RIE). The MVG consists of two properly insulated electrodes integrated on top of an atomic force microscopy (AFM) tip, forming a coaxial embodiment. The special design enables us to vary the cathode-anode separation and the turn-on voltage changes accordingly. The experiments show that the MVGs operate at low bias potential and demonstrate very good I-P dependence over a wide pressure range.

1. INTRODUCTION

Vacuum metrology in the high and ultra-high vacuum regime plays an important role in modern industry as well as scientific research. Most common vacuum gauges used in these pressure regimes rely on ionization gauges [1, 2]. These gauges measure the electrical current resulting from the collection of the molecules ionized by electrons accelerated in high electric fields inside the gauge, which is used as an indirect measure of gas density and thus pressure [3, 4]. Device-dependent features like response time, power consumption, outgassing heat and light emission, or limited filament life time must be considered when choosing the proper vacuum gauge type for the respective application [5].

To reduce the power consumption of cold-cathode ion gauges, high-aspect-ratio cathodes were fabricated resulting in a high field-enhancement factor [6–7]. However, small fabrication variations, e.g., in emitter tip radius, height, chemical composition or crystalline orientation, can result in huge variations in emission current. Additionally, the strong electric field gradient near the cathode could cause the adsorption of gas molecules from the surroundings, thereby changing the local work function and again altering the emission efficiency.

2. EXPERIMENT

In this paper, we present an approach for the reproducible integration of an individual field-emitter device into a Si-AFM tip utilizing FIB processing and microfabrication techniques. Figure 1 shows the principle sketch and the fabrication scheme of the MVG.

First, a 5 nm thick titanium (Ti) adhesion layer followed by an 80 nm thick gold (Au) and a 5 nm thick Ti layer are sputtered onto the Si-AFM cantilever. The Au layer has a later role as the field emitter (FE) - the cathode. The uppermost Ti layer provides better adhesion between the Au layer and the subsequently-deposited about 600 nm thick Si_3N_4 insulation layer, deposited by plasma enhanced chemical vapour deposition at 300°C. In order to improve the insulation capability, an additional 40 nm thick ZrO_2layer is deposited by atomic layer deposition.

On top of the insulation layer, another 5 nm thick Ti adhesion layer followed by an 80 nm thick Au layer are sputter deposited, later forming the ring-shaped anode around the tip. Finally, before the tip exposure by FIB cutting and RIE the probes are sputter-coated with a 50 nm thick

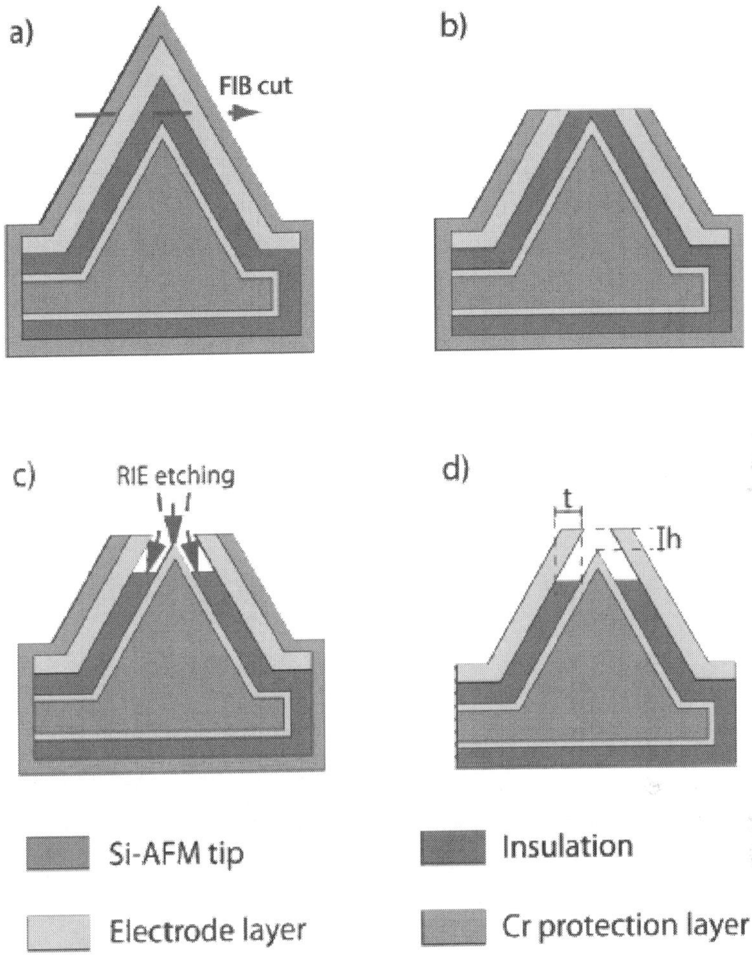

Figure 1. Process flow chart for MVG fabrication: (a) AFM tip with conductive Ti/Au/Ti layers, 600 nm Si_3N_4 and 40 nm ZrO_2 as insulating layers and a 50 nm Cr protection layer; (b) tip after FIB cutting of the insulation from the apex of the tip; (c) gap formation by selective ICP-RIE etching of insulation layer; (d) fully featured MVG after the removal of the Cr layer by wet etching. Insulation thickness t and vertical anode/cathode offset h implicate the MVG geometry modulation sites, where the electrode separation is controlled by varying the insulation layer thickness or position of the FIB cutting

sacrificial Cr protection layer. The final MVG blank with the completed layered structure is shown in Figure 1a. Afterwards, the Cr, Au and insulation layers are partly removed at the very apex of the tip with the FIB, utilizing a beam of Ga^+ions at 30 keV and a beam current of 50 pA (Figure 1b). By careful adjustment of the FIB milling, the innermost conductive tip is kept buried underneath a Si_3N_4 insulation layer about 150 nm thick. Subsequently, an inductively-coupled-plasma RIE process is applied in two steps. First, a mixture of $SiCl_4$ and SF_6 is required for removal of the Ga-containing top insulation layer. Previous experiments have shown that Ga arising from the FIB processing acts as an etch-stop during Si_3N_4 dry etching with SF_6 [8, 9]. In the subsequent RIE etching step, only SF_6 is used to etch Si_3N_4selectively, exposing the conductive tip, whereby the Cr layer protects the rest of the sample as shown in Figure 1c. The removal of the insulation layers results in a gap between the conical cathode tip and the surrounding ring-shaped anode. After successful electrode exposure, the protective Cr layer is removed by wet etching with Microposit chrome etch 18. Figure 2 shows the SEM image of the fully featured MVG.

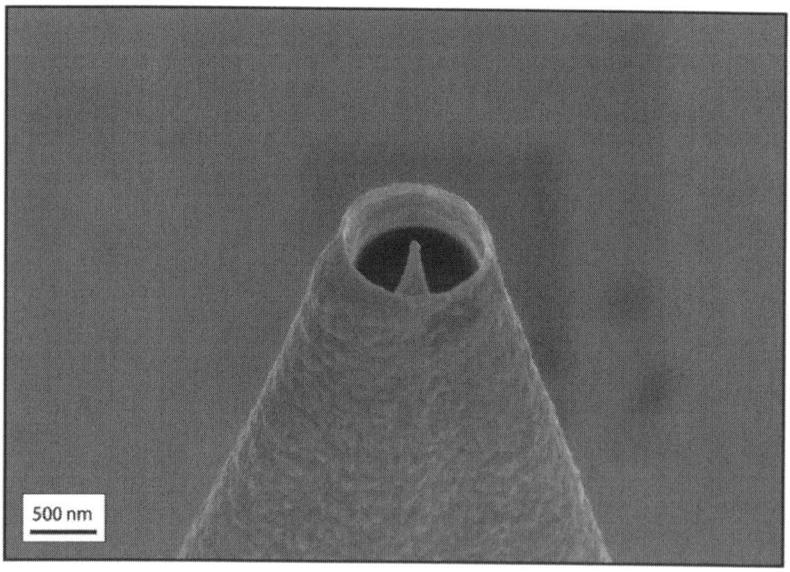

Figure 2.SEM image of the MVG showing the Au cathode with an apex diameter of about 60 nm, surrounded by the ring-shaped Au anode with an in-between distance of roughly 640 nm

For this particular tip, the Au cathode is approximately 1µm long and has an apex diameter of about 60 nm. As indicated in Figure 1d, our MVG processing approach allows easy modulation of the MVG electrode separation by varying the insulation layer thickness (t) or the vertical anode/cathode offset (h) by tuning the position of the FIB cut.

Overall, we fabricated more than 30 devices for field emission as well as electrochemical investigations [10], demonstrating the versatility of this approach. With respect to the miniaturized vacuum gauges, we fabricated four devices with a cathode-anode spacing of about h=500 nm, and two devices with a spacing of about 1 µm. To investigate long-term stability we tested the devices for 8 h. Thereby all of this time the vacuum gauges appeared stable and the pressure dependency measurements were well reproducible.

3. RESULTS AND DISCUSSION

In order to investigate the field-emission (FE) properties and the pressure dependency, the wire-bonded MVG was glued on a chip carrier and mounted in a vacuum chamber. Once the base pressure of 5.7×10^{-7} mbar was reached, the emission-current characteristic was investigated as a function of the applied potential. The pressure was gradually increased, regulated dynamically by a leak valve, and for each pressure value an I-V characteristic was recorded. Figure 3 shows the pressure-dependent I-V characteristics of a MVG with a cathode-anode spacing of h_1=640 nm as a function of the pressure. Remarkably, the MVG shows a very low threshold voltage of about 2 V and a high FE current of 300 µA for an applied voltage of 10 V at a base pressure of about 5.7×10^{-7} mbar.

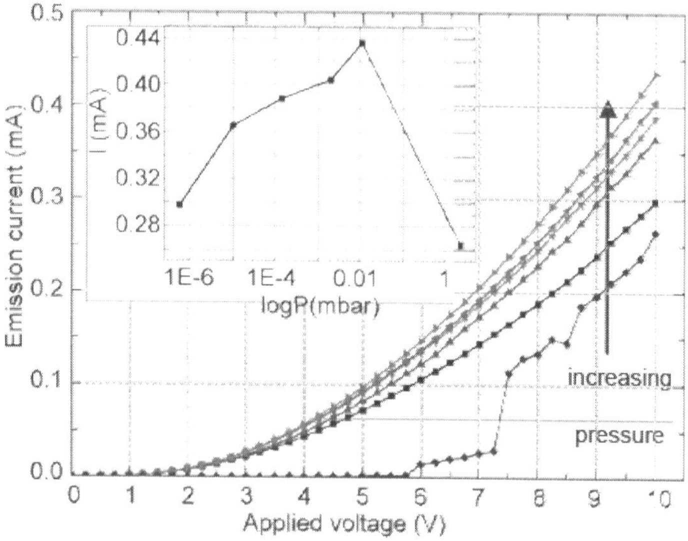

Figure 3. Emission current of the MVG versus the applied bias at vacuum pressures of 2.3 mbar, 1.1×10^{-2} mbar, 2×10^{-3} mbar, $1{,}4\times10^{-4}$ mbar, 1×10^{-5} mbar and 5.7×10^{-7} mbar; the respective inset shows the corresponding I-P characteristics for an applied bias of 10 V

As expected, the FE current increases with the pressure up to about 1.1×10^{-1} mbar. The I-P characteristic for a potential of 10 V in the inset of Figure 3 shows a nearly linear characteristic in the wide pressure range from 5.7×10^{-7} to 1.1×10^{-2} mbar. This behaviour was observed to be reproducible and time stable for at least 8 h. For even higher pressures an abrupt decrease in the FE current and unstable operation was observed.

Furthermore the influence of the cathode-anode separation was investigated by comparing the emission properties of the above-mentioned tip (h_1=640 nm) and another MVG with a much larger electrode spacing of h_2=1.3 μm. As shown in Figure 4a, the increase of the cathode-anode separation results in an increased turn-on voltage of 35 V and a significantly lowered emission current.

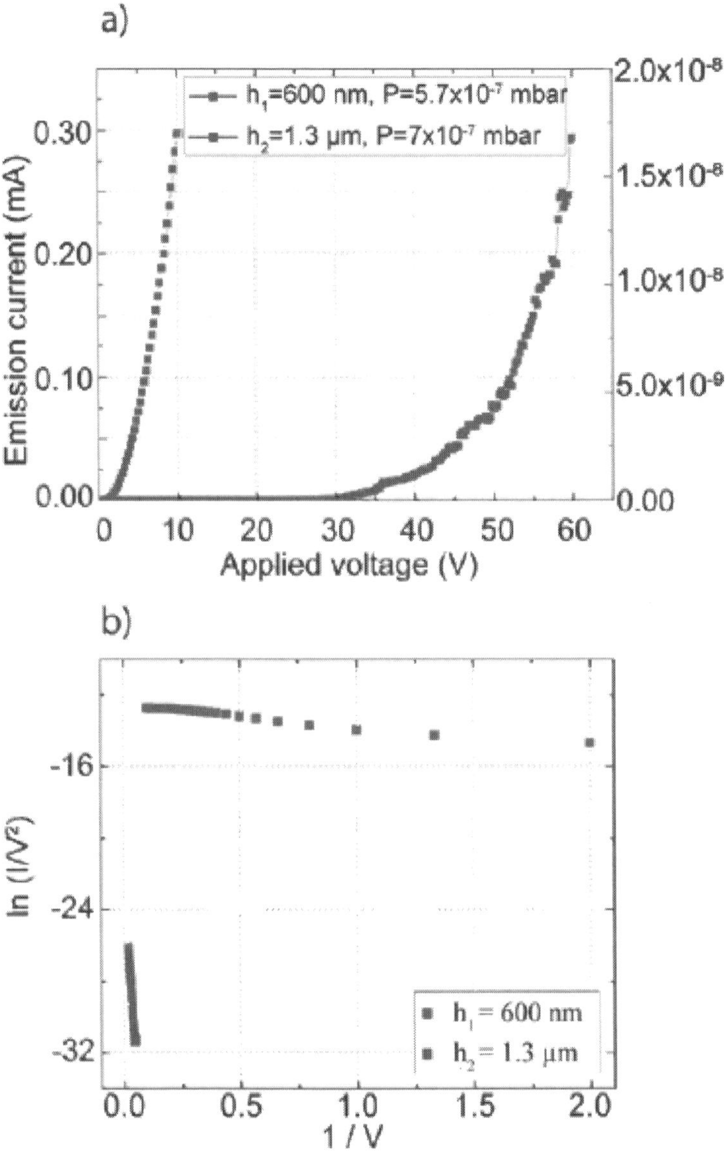

Figure 4. (a) Comparison of the I-V characteristic of a MVG with cathode-anode separation of h_1~640 nm (red) and h_2 ~1.3 μm (blue); (b) the respective F-N plots

In order to qualitatively investigate the FE behaviour, the I-V characteristic was plotted according to the FN theory [11], as shown in the respective F-N plot in Figure 4b. The turn-on voltage dependency of the cathode anode separation is best displayed by the increase of the F-N slope with increasing h. The F-N plots are also used to obtain the characteristic field-enhancement factor [12]. For the calculation of the current density of a field-emission device, it is assumed that the electrode surface is planar and the potential within the electrodes constant. For a planar electrode, the electric field is given by E=V/h, where V is the applied voltage and h the distance between the cathode and the anode.

If the cathode has a non-planar structure (i.e., a conical shape), the resulting field non-linearity has to be considered in the calculations. The maximum electric field will be found at the apex of the cathode tip due to the compression of the equipotential surfaces around the cathode. Although the exact calculation of the field-emission current for a sharp cathode requires the numerical calculation of three-dimensional Poisson and Schrödinger equations for electron emission, it is convenient and commonly accepted to introduce a geometry-dependent correction factor given by:

$$E = \beta V$$

where β is the field-enhancement factor calculated from:

$$\beta = \gamma / h$$

with γ the geometric enhancement factor and h the cathode-diode separation.

Assuming the work function for Au to be Φ=4.8 eV [13, 14] the field-enhancement factor β [15] for our samples was calculated to about 4.2×10^{7} m^{-1}. Even though our geometry deviates from that of the ideal high-aspect-ratio cathode in combination with a flat anode this huge "field-enhancement factor" demonstrates that the nanomachined, cone-shaped cathode and the ring-shaped anode acts as a strong field-enhancement embodiment, enabling a low turn-on voltage of the miniaturized field-emission vacuum gauge.

4. SUMMARY AND CONCLUSIONS

In summary, we have shown an alternative self-aligned fabrication approach for miniaturized cold-cathode vacuum gauges on top of Si-AFM tips. The experiments showed that optimized MVGs operate with very low bias potential and show very good I-P dependence over a remarkably wide pressure range between 5.7×10^{-7} and 1.1×10^{-2} mbar. By increasing the cathode-anode separation, the turn-on voltage changed accordingly.

The unique geometry of the MVG allows wide-range pressure measurements in systems that require vacuum gauges to operate at low bias potential. The simple structure, compact size and particular embodiment of this MVG also enables spatially resolved pressure measurements, e.g., monitoring the inflow of gases from a nozzle into a vacuum chamber.

5. ACKNOWLEDGEMENTS

This work was supported by the Austrian Research Promotion Agency (Die Österreichische Forschungsförder-ungsgesellschaft mbH, FFG) and the government of Lower Austria through the COMET programme, as well as the Austrian Society for Micro-and Nanoelectronics (GMe). Furthermore, the authors thank the Centre for Micro-and Nanostructures (ZMNS) for providing the clean-room facilities.

- Received March 20, 2014.
- Accepted October 10, 2014.

REFERENCES

1. Wilfer S, Edelmann C (2004) Miniaturized vacuum gauges. Vac. Sci. Technol. A 22: 309-320.
2. Górecka-Drzazga A (2009) Miniature and MEMStype vacuum sensors and pumps. Vacuum 83(12): 1419-1426.
3. Leck J H (1957) Pressure measurement in vacuum systems. Second edition. London: Science Paper- backs, Chapman and Hall.
4. Wu H Y, Yang T N, Wang X J (2013) Research on Pressure Measurement Method in Vacuum Envi- ronment. Applied Mechanics and Materials 19: 347-350.
5. Wilfert S, Edelmann C (2012) Field emitter based vacuum sensor. Vacuum 86: 556-571
6. Yang Y C, Liu L, Jiang K L, Fan S S (2012) A newly developed vacuum sensor using carbon nanotube (CNT) field-emission has been designed. Vacuum 86: 885-888.
7. Su J, Guo D Z, Zhang G M (2012) Miniature coldca- thode ionization gauge based on composite films of carbon nanotubes and MgO nanoparticles. 25th Vacuum Nanoelectronics Conference (IVNC), 1-2.
8. Chekurov N, Grigoras K, Peltonen A, Franssila S, Tittonen I (2009) The fabrication of silicon nano- structures by local gallium implantation and cryogenic deep reactive ion etching. Nanotechnol- ogy 20: 065307.
9. Schmidt B, Oswald S, Bischoff L (2005) Etch rate retardation of Ga+ion beam irradiated silicon. J. Electrochem. Soc. 152: G875-G879.
10. Avdic A, Lugstein A, Wu M, Gollas B., Pobelov I, Wandlowski, Leonhardt K., Denuault G, Bertagnol- li E (2011) Fabrication of cone-shaped boron doped diamond and gold nanoelectrodes for AFM–SECM. Nanotechnology 22: 145306.
11. Fowler R.H, Nordheim L (1928) Electron Emission in Intense Electric Fields. Proc. R. Soc. Lond. A 119: 173-181

12. Lewis P A, Alphenaar B W, Ahmed H (2001) Measurements of geometric enhancement factors for silicon nanopillar cathodes using a scanning tunneling microscope. Appl. Phys. Lett. 79: 1348-1350.

13. Anderson P A (1959) Work function of gold. Phys. Rev. 115: 553.

14. Hiroyuki K (2008) Effective work functions for ionic and electronic emissions from mono-and polycrys- talline surfaces. Progress in Surface Science 83(1): 1.

15. Smith R C, Forrest R D, Carey J D, Hsu W K, Silva S R P (2005) Interpretation of enhancement factor in nonplanar field emitters. Applied Physics Letters 87: 013111.

8

VARIABILITY AND RELIABILITY OF SINGLE-WALLED CARBON NANOTUBE FIELD EFFECT TRANSISTORS

Ahmad Ehteshamul Islam [1,2]

[1]Soft Matter Materials Branch, Materials and Manufacturing Directorate, Air Force Research Laboratory, Wright-Patterson Air Force Base, Dayton, OH 45433, USA

[2]Biological and Nanoscale Technologies, UES Inc., Dayton, OH 45432, USA

ABSTRACT

Excellent electrical performance and extreme sensitivity to chemical species in semiconducting Single-Walled Carbon NanoTubes (s-SWCNTs) motivated the study of using them to replace silicon as a next generation field effect transistor (FET) for electronic, optoelectronic, and biological applications. In addition, use of SWCNTs in the recently studied flexible electronics appears more promising because of SWCNTs' inherent flexibility and superior electrical performance over silicon-based materials. All these applications require SWCNT-FETs to have a wafer-scale uniform and reliable performance over time to a level that is at least comparable with the currently used silicon-based nanoscale FETs. Due to similarity in device configuration and its operation, SWCNT-FET inherits most of the variability and reliability concerns of silicon-based FETs, namely the ones originating from line edge roughness, metal work-function variation, oxide defects, *etc.* Additional challenges arise from the lack of chirality control in as-grown and post-processed SWCNTs and also from the presence of unstable hydroxyl (–OH) groups near the interface of SWCNT and dielectric. In this review article, we discuss these variability and reliability origins in SWCNT-FETs. Proposed solutions for mitigating each of these sources are presented and a future perspective is provided in general, which are required for commercial use of SWCNT-FETs in future nanoelectronic applications.

Keywords: single-walled carbon nanotube; field effect transistor; variability; reliability; hydroxyl group passivation; oxide defect; hysteresis; noise; radiation dose; degradation

1. INTRODUCTION

Single-Walled Carbon NanoTubes (SWCNTs) were first discovered in 1991 [1] and their electronic properties depend on the directions at which two-dimensional graphene stripes are (hypothetically) rolled to make SWCNTs [2,3,4]. When a SWCNT is defined by rolling a graphene layer and by joining any of the solid symbols of Figure 1a with the lattice origin $(a_1,a_2) = (0,0)$, resultant SWCNT will have metallic properties with no bandgap in the energy *vs.* density of the state's relationship (Figure 1b). On the other hand, SWCNTs defined by joining the open symbols of Figure 1a with the lattice origin will have semiconducting properties with bandgaps (Figure 1c,d) that can be expressed as [2,3]:

$$E_G = \frac{2a_{C-C}|t|}{d}. \tag{1}$$

In Equation (1), d is the diameter of SWCNT and equals $\sqrt{3}a_{C-C}/\pi \times \sqrt{(n^2 + nm + m^2)}$, (n,m) is the chiral index designating the distance between the open symbols and the lattice origin via chiral vector $\bar{c} = n\bar{a}_1 + m\bar{a}_2$ (Figure 1a), $a_{C-C} = 0.142$ nm is the C–C bond length, $t \sim -3$ eV is the C–C bonding energy. SWCNTs with $n = m$ are called armchair nanotubes and always show metallic behavior, whereas SWCNTs with $n \neq m$ either show metallic behavior for mod $(n - m)/3 = 0$ or semiconducting behavior for mod $(n - m)/3 = 1, 2$. SWCNTs with $m = 0$ are called zigzag nanotubes and the rest with $n \neq m \neq 0$ are called chiral nanotubes.

Semiconducting SWCNTs (s-SWCNTs) are of particular interest for future electronic devices, where s-SWCNTs act as a channel in conventional field effect transistor (FET) configuration. Carrier mobility in such s-SWCNT-FETs can be as high as ~10^5 cm^2/V-s [5,6], which outperforms similar values in current-day FETs. In addition, the ultra-thin s-SWCNT layer that forms the channel of FET offers better electrostatic gate control, hence less short channel effect compared to classical silicon-based FETs [7]. Such ultra-thin layers also offer extreme mechanical flexibility to the resultant electronic device, thereby promoting the best platform for future flexible electronic devices [8,9], which are currently being manufactured using FETs made on thin silicon substrates [10,11]. Use of s-SWCNTs as channels, m-SWCNTs as contacts [12] and memory elements [13], SWCNT network [14,15] or single s-SWCNT [16,17] as memory, and CNT fabrics as coaxial cable [18] may open up the possibility of having "all-CNT" electronics for high performance applications. In addition to electronics, SWCNT-FETs are also considered as chemical-/bio-sensors because of their higher surface-to-volume ratio that reduces response time [19] and extreme sensitivity [20,21], and their selectivity [20] to chemical and biological species.

To enable all these applications stated in the last paragraph, SWCNT-FETs and memories not only require superior performance, but also require a controlled level of variability and reliability. Such controllability has to ensure that performance of associated devices does not drop below a certain limit, which can be compensated through circuit design via guard bands [22,23]. Figure 2 schematically explains the extent of guard band that is used in circuit design for handling variability and reliability. Devices have different variability sources that give rise to time-zero variation in device performance around its mean value. In addition, during the operation of the device, its performance degrades over time and such degradation comes with its own statistical variation. Based on this knowledge, circuits are designed with such devices considering that the tail of the performance distribution at the end of intended lifetime will not go beyond a certain limit—thus, the designed circuits are guard-banded against performance variations. As device performance improves via Moore's law scaling [24], device design is continuously modified to minimize guard-band such that the devices can be operated at higher performance without compromising the lifetime.

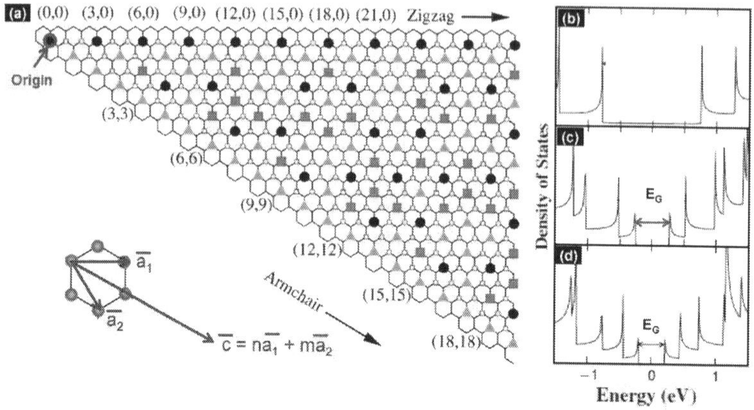

Figure 1. (**a**) Two-dimensional graphene lattice for defining structures of (*n*,*m*) Single-Walled Carbon NanoTubes (SWCNTs) (reprinted with permission from Ref. [3]; © 2005 by the American Physical Society.). Definition of lattice vectors a_1, a_2 with respect to carbon atoms in the graphene lattice is shown in the bottom part of the figure. SWCNTs can be formed by rolling this sheet to join the origin (a_1,a_2) = (0,0) with other points. Joining the open symbols with the origin makes semiconducting SWCNTs, while joining the solid symbols with the origin makes metallic SWCNTs. (**b**) Density of states (DOS) *vs.* energy plot obtained using a tight binding model [4] for an armchair (10, 10) SWCNT having metallic property. (**c**,**d**) DOS + energy plot for a chiral (11, 9) SWCNT and a zigzag (22, 0) SWCNT. Both these SWCNTs have semiconducting properties and a chirality-dependent band gap (E_G) defined using Equation (1). (**b**–**d**) are reprinted with permission from Ref. [4]; © 2003 IOP Publishing Ltd and Deutsche Physikalische Gesellschaft, Published under a CC BY-NC-SA license.

In this review article, we will discuss different aspects of variability and reliability in SWCNT-FETs that will eventually define guard-band in such devices. Section 2 discusses the configurations and performance of SWCNT-FETs that has potential use in future electronics. Section 3 highlights the sources of variability in SWCNT-FETs and the current solutions to control them. Section 4 performs a similar review for the reliability sources. Section 5 summarizes the paper and provides brief guidelines for improving variability and reliability phenomena in SWCNT-FETs based on current understanding of CNT growth conditions and device fabrication challenges.

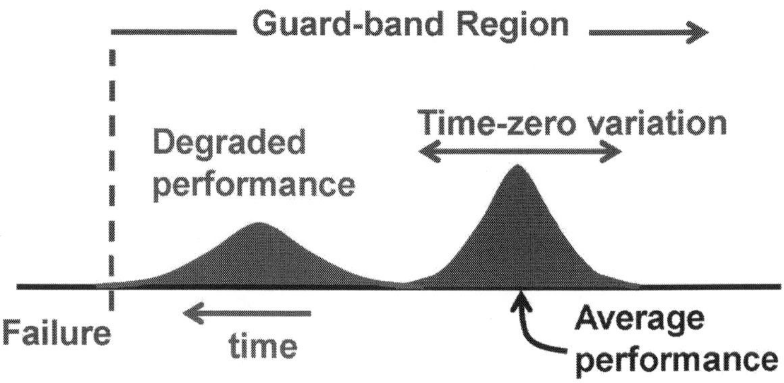

Figure 2. Schematic representation of the amount of guard-band that is used in electronics for handling process (or time-zero) variation and reliability (time dependent variation) in device performance. The performance of a device has a distribution over and above its mean value and it degrades over time. Once of the tail of performance distribution for the device reaches the guard-band margin, the associated device is designed to fail.

2. CONFIGURATIONS AND PERFORMANCE OF SWCNT-FET

2.1. Device Geometry

For FET applications, SWCNTs are commonly grown in random networks or aligned arrays using metal catalyst placed on SiO_2 [9,12,25,26]. Alternately, SWCNTs are drop-casted from solution on SiO_2 [27,28,29] or hafnium dioxide (HfO_2) [30,31] and assembled in random networks [27,28,29] or parallel arrays [31,32] via coating of self-assembled monolayers on oxides. Bottom-gated FETs with thick SiO_2 as gate dielectric and heavily doped silicon as gate are widely used test configurations for SWCNT-FETs. Such bottom-gated structures though are suitable for bio-/chemical-sensing applications [20,21,33], top-gated geometry with thin, high-κ gate dielectric is preferred for high performance electronics [34,35]. (In some novel SWCNT-FETs, local bottom-gating [36], lateral gating with suspended SWCNTs [37], and surround-gate geometries [38] are also used for high performance electronics.) Measurement of SWCNT properties (e.g., conductivity, mobility, contact resistance of individual SWCNTs and their diameter dependence) by applying suitable bias at source, drain, and gate terminals involves fabrication of single SWCNT-FET. Schematic of such a FET in dual-gate (*i.e.*, having top and bottom gate with suitable gate oxides) configuration is shown in Figure 3a. Such single SWCNT-FET, however, cannot provide the amount of current required for different applications, e.g., ~mA per µm of channel width is required for high performance electronics [35]. Current in SWCNT-FETs can be increased either by having a network (Figure 3b) or an aligned array (Figure 3c) of SWCNTs' bridging sources and drain terminals. Magnitude of current in these configurations depends on the density of SWCNTs (per unit area for network SWCNTs of Figure 3b or per unit width across the length of SWCNTs for aligned SWCNTs of Figure 3c), diameter of SWCNTs, chirality of SWCNTs, and electrostatic coupling on SWCNTs via gate. For network SWCNT-FETs, current additionally depends on the charge transfer efficiency across different SWCNTs in the network [39,40,41].

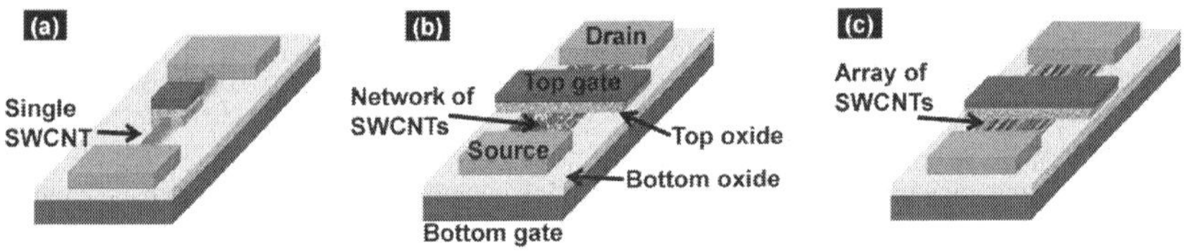

Figure 3. Schematic representation of SWCNT-based field effect transistor (FETs) in top- and bottom-gated configuration for (**a**) single SWCNT, (**b**) SWCNT networks, and (**c**) aligned array SWCNTs. Generally, SWCNTs are grown or coated on oxides and then the source, drain, top oxide and top gate are formed on top of them via lithography. Though the schematics here show partial coverage of SWCNTs in between source and drain, the actual top-gate SWCNT-FET application uses full coverage of SWCNTs.

2.2. Materials for SWCNT-FETs

As mentioned before, SWCNTs are grown on insulating substrates like SiO_2, quartz, sapphire, *etc.* using metal (e.g., Fe, Ni, Co and their alloys) nanoparticles as catalyst. Length, diameter distribution, and chirality of grown SWCNTs vary among the growth methods being used [42,43]. In general, for as-grown SWCNTs, length varies from ~1 µm to several 100 µm, diameter ranges between ~0.7 nm and ~5 nm

[26,44,45,46], density is approximately 14–200 per μm^2 for network SWCNTs [12,31] and ~2–45 per μm across the length of the SWCNT for aligned SWCNTs [26,31,44,47,48], and chirality varies randomly. Long SWCNTs (>100 μm) are only obtained during chemical vapor deposition (CVD) on stable temperature- (ST-) cut quartz substrates [25,49]. Diameter and chirality of as-grown SWCNTs can be sorted by dissolving SWCNTs in surfactants and then using post-processing techniques like ultra-centrifugation [50], gel electrophoresis [51], chromatography [52]. Such solution-processed, sorted SWCNTs can be assembled into a high-density compact film via Langmuir-Blodgett method [35]. However, solution-processing breaks SWCNTs into small segments (~μm or less) and reduces the carrier mobility or conductivity of SWCNTs (down to ~1–20 cm^2/V-sec [35,46] with a maximum of ~200–300 cm^2/V-sec [28,53]). An alternative approach of maintaining the integrity of SWCNTs in terms of length (up to several hundreds of μm) and carrier mobility (>1000 cm^2/V-sec) is to coat horizontally aligned SWCNTs with a molecular film and use thermo-lithography (thermal analogous of photo-lithography) to etch-off m-SWCNTs from the substrate [54]. Remaining s-SWCNTs have a low density; however, there is potential for density improvement via multiple transfers [55,56].

In addition to the requirement of having substrates with high density s-SWCNTs, future electronic devices require appropriate doping in s-SWCNTs to control their conduction. Absence of molecular doping in SWCNTs, however, necessitates the use of metals with lower (higher) work-functions to conduct electrons (holes) through the SWCNTs, hence make n- (p-) type SWCNT-FETs. Palladium (Pd) with work-function of ~5.1 eV is generally used as source/drain contacts for making p-type SWCNT-FETs that allows hole flow from contacts into and out of the SWCNT via Schottky contacts for $d < 1$ nm SWCNTs and via ohmic contact for $d > 1$ nm SWCNTs [57,58,59]. Presence of negatively charged hydroxyl (-OH) groups at SWCNTs interface also makes the p-type conduction in SWCNT-FETs easier. N-type or electron conduction in SWCNTs, on the other hand, can be enabled using calcium (Ca) and scandium (Sc) source/drain contacts with work-functions ~3–4 eV [60,61]. In such case, conduction can be further enhanced via electrostatic doping with high-κ oxides [38,62].

2.3. FETs with Single SWCNT

Current-voltage (I-V) characteristics of FETs with single s-SWCNT have reasonably good switching behavior in the measured transfer characteristics (*i.e.*, drain current (I_{ds}) *vs.* gate voltage (V_g) characteristics, as shown in Figure 4a) and good saturation in output characteristics (*i.e.*, I_{ds} *vs.* drain to source voltage (V_{ds}) characteristics, as shown in Figure 4b) for any channel length (L_{ch}) down to ~10 nm [63]. These FETs have predominant p-type behavior because of the natural presence of –OH groups at the interface of SWCNTs. On-current (I_{on}; defined as I_{ds} at a certain high $|V_g - V_t|$, where V_t is the threshold voltage of a FET) and off-current (I_{off}; defined as the minimum of I_{ds}) in these FETs with single s-SWCNT show variation with d and source/drain contact material being used. Contacts with higher work-function (Φ_{WF}) can inject holes into the valence bands of s-SWCNTs more easily and have larger I_{on} and I_{off} (Figure 4c,d); [44,58,59]), especially for larger diameter s-SWCNTs that have smaller band-gap, E_g (see Equation (1)) and hence have negligible hole injection barrier at the metal/s-SWCNT interface. Therefore, contacts play a major role in the conduction of SWCNT-FETs and are considered to have dominant contribution in limiting I_{ds} in the near ballistic limit [28,63], when L_{ch} is much less than the channel mean free path that defines mobility. Sub-threshold slope (SS; defined as $\partial V_g/\partial(log_{10}I_{ds})$ in the region with $V_g << V_t$) for FETs with single s-SWCNT varies from ~80–95 mV/dec and increases with a decrease in L_{ch} due to ambipolar conduction [64]. (Use of appropriate chemical coating on SWCNT in a dual-gate device configuration can reduce SS to 63 mV/dec [65], which is close to the theoretical limit of $SS = 2.3k_BT/q \sim$ 60 mV/dec at room temperature for a FET [66]; here, k_B is the Boltzmann constant, T is temperature in °K, and q is the electron charge.) Hole effective mobility ($\mu_{eff,h}$) of s-SWNTs ranges from ~10^3–10^5 cm^2/V-sec [5,6,25,63,67] and increases with d [5]. Such $\mu_{eff,h}$ value depends on CNT growth conditions and is better for CVD-grown CNTs compared to solution-processed CNTs [28].

In contrary to s-SWNTs, FETs with single metallic (m-) SWCNT-FETs have very small gate modulation and show ambipolar behavior with a smaller n-type tail (Figure 5a). Such gate modulation is often attributed to the presence of Mott-insulating states [68] and/or strain-induced bandgap [69] in m-SWCNTs. FETs with s-SWCNTs also show ambipolar behavior, more significantly at higher $|V_{DS}|$ and for larger diameter s-SWCNTs (Figure 5b) due to band-to-band tunneling (BTBT) [70].

2.4. FETs with Multiple SWCNTs

Multiple SWCNTs in FET's channel are either arranged randomly as network or aligned parallel to each other bridging source and drain. Network SWCNT-FETs can be made using CVD-grown SWCNTs with mixed chiralities (Figure 6a shows a scanning electron microscope (SEM) image for one such bottom-gated FET near a source/drain contact). These FETs have low on/off ratio, unless the conduction through metallic SWCNTs is turned off via periodic stripping [9]. An alternate way of improving on/off ratio is to use solution-processed, chirality sorted s-SWCNTs for making FETs. However, such FETs show lower conductance compared to CVD-grown SWCNTs because of the presence of structural defects [28].

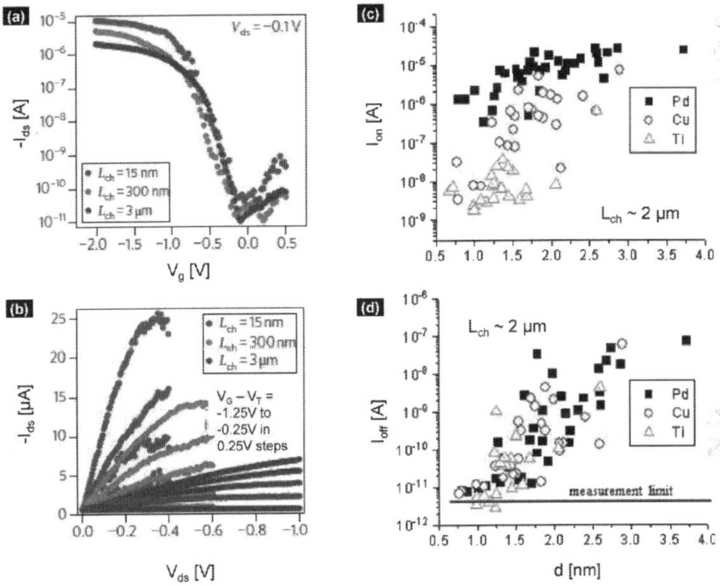

Figure 4. (**a**) Transfer or I_{ds}-V_g and (**b**) output or I_{ds}-V_{ds} characteristics of single s-SWCNT-FETs (reprinted by permission from Macmillan Publishers Ltd: Nature Nanotechnology [63], © 2010). Diameter dependence of (**c**) on-current I_{on} and (**d**) off-current I_{off} for s-SWCNT-FETs using different types of metals (palladium: Pd, copper: Cu, titanium: Ti) as source/drain (reprinted with permission from Ref. [58], © 2006 American Chemical Society.). In general, easier hole conduction for higher work-function (Φ_{WF}) contact metals like Pd results higher $I_{on}$and I_{off} for associated FETs.

Network SWCNTs have impedances at the junctions of different SWCNTs; consequently, they have lower conductance and are suitable only for applications like non-volatile memories [14,15]. Parallel arrangement of aligned SWCNTs resolves the issue with junction impedance and promises applications of SWCNTs in high performance radio frequency (RF) electronics [47,71,72], and transparent electronics [73,74]. Figure 6b shows a SEM image of one such top-gated FET before gate dielectric and gate metal formation. These aligned array SWCNT-FETs are either fabricated by growing SWCNTs via CVD on special types of quartz or sapphire substrates, whose crystal orientations [49] and/or step-edges [75,76] help parallel alignment (with less than 0.1% imperfection [77]) of $d \sim$ 0.7~3 nm SWCNTs [26,44,45], or fabricated by aligning solution-processed SWCNTs via dielectrophoresis [46]. CVD-grown SWCNTs are long (with a length of several hundreds of μm [77]) and have high conductivity [5,25,63,67], however, require the use of post-processing techniques like thermo-lithography [54] and laser irradiation [78] for removal of m-SWNTs, or electrical breakdown [79] for breaking the continuity of m-SWNTs, and hence, cause FETs to use the remaining s-SWNTs.

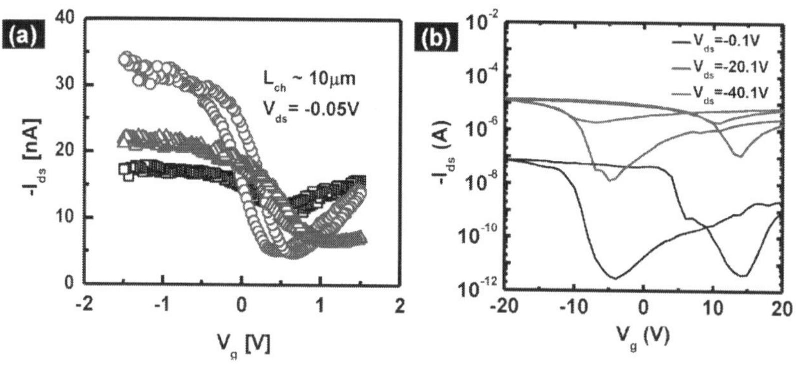

Figure 5. (**a**) Transfer characteristics of single metallic (m-) SWCNT-FETs showing ambipolar behavior (reprinted with permission from Ref. [44]; © 2012, AIP Publishing LLC). (**b**) Ambipolar behavior is also observed for FETs with a $d \sim$ 1.7 nm s-SWCNT and at larger V_{ds}(reprinted with permission from Macmillan Publishers Ltd: Nature Nanotechnology [54], © 2013).

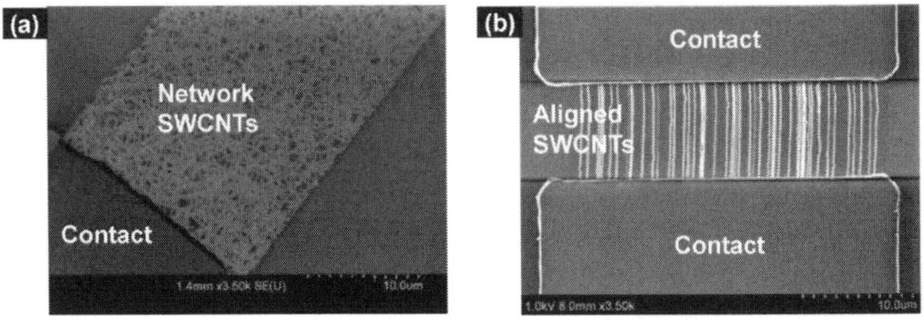

Figure 6. Scanning electron micrograph (SEM) of (**a**) random network of SWCNTs near one of the source/drain contact and (**b**) horizontally aligned SWCNTs bridging source/drain contacts.

3. VARIABILITY IN SWCNT-FETS

3.1. Variability in FETs with Single SWCNT

Performance variation in FETs with single s-SWCNT arises from the variation in d, contact and gate metals' Φ_{WF}, L_{ch}, T_{ox} (oxide thickness), and interface and oxide defects from one FET to another. For FETs with L_{ch}>μm, variation in d can give rise to several orders of magnitude change in I_{ds}, as shown in Figure 7a for small V_{ds} = −50 mV. One can simulate such transfer characteristics for different d (dashed lines in Figure 7a) by considering the conductance of the FET (G_{ds}) as a series combination of conductances of s-SWCNT (G_{ss}; see Ref. [44] for appropriate expressions) and contact ($G_c = G_{c0}T_c$; where T_c is the transmission probability of carriers near the source/drain contact) with components from both electron (e) and hole (h), *i.e.*,

$$G_{ds} = G_{ds,e} + G_{ds,h} = \left(G_{ss,e}^{-1} + G_{c,e}^{-1}\right)^{-1} + \left(G_{ss,h}^{-1} + G_{c,h}^{-1}\right)^{-1}, \quad (2)$$

and hence, calculate I_{ds} using $I_{ds} = G_{ds} \times V_{ds}$. Simulated results for 0.6 nm < d < 1.75 nm are consistent with measurement at $V_g < V_t$, where G_{ss} dominates G_{ds}. The differences in the region of $V_g > V_t$, where G_c dominates G_{ds}, partly reflect variation in the fixed component of G_c [80] that is not considered in simulation and also reflect I_{ds} measurement limitation set at ~pA. Plots of diameter dependence of I_{on} ($\equiv I_{ds}$ @ $V_g - V_t$ = −1V, V_{ds} = 50 mV), maximum transconductance $G_{m,max}$ ($\equiv$max $|\partial I_{ds}/\partial V_g|$), and V_t (Figure 7b) provide additional detail about the origin of performance variation in s-SWCNT FET. At large diameters, when the transmission of carriers through the Schottky barrier near the contacts is unity, variation of $I_{on} \sim d$ and $G_{m,max} \sim d^2$ originates from the diameter dependence of μ_{eff} in s-SWNTs [5,81,82]. Similarly, variation of V_t($\equiv V_g$ @ $|I_{ds}|$ = $|I_{ds,max}|/100$) with d follows that of I_{on}. At small diameters, non-linear dependence of transmission through the contact Schottky barrier leads to a non-linear variation in I_{on}, $G_{m,max}$, and V_t with d.

Simulated I_{on}, $G_{m,max}$, and V_t distributions (calculated using diameter distribution of Figure 7c and variations of I_{on}, $G_{m,max}$, and V_t *vs.* d from Figure 7b) suggest good consistency with respective measurements (Figure 7d–f). Similar to the diameter distribution of Figure 7c, I_{on} distribution fits log-normal statistics, except near the lower tails of the distribution where transport through the contacts' Schottky barrier with $T_c < 1$ dominates I_{ds}. Measured V_t − $<V_t>$ distribution (where $<V_t>$ is the average of the distribution) is wider than the simulated one due to the extra contributions from defects [16,83] and gate Φ_{WF} [84,85] variations that is not considered in the simulation. Passivation of SWCNT interface using hydrophobic self-assembled monolayers (SAMs) can tighten the V_t distribution by reducing hydroxyl (–OH) group related interface defects [86]. Figure 8a demonstrates utility of such SAM-based approach for reducing V_t distribution, where L_{ch} << μm s-SWCNT FETs are made using solution-processed SWCNTs and passivated with hexamethyldisilazane (HMDS). As in CVD-grown s-SWCNTs, distribution in G_m (Figure 8b) for such solution-processed s-SWCNT FETs follows that of diameter (Figure 8c). In addition, contact resistance $R_c = 1/G_c$ in such small L_{ch} s-SWCNT FETs plays a dominant role in transport. Distribution of R_c(Figure 8d) arises from variations in T_c with d [28] and also from variations in the contact's Φ_{WF}. In contrary, FETs with m-SWCNT and similar L_{ch} have tighter R_c distribution [28]; however, they still suffer from variations in I_{ds} (Figure 8e) and, hence, G_m (Figure 8f). Since m-SWCNTs are unattractive for high performance electronics applications requiring high I_{on}/I_{off}, parametric variations in m-SWCNT FETs are not well studied.

3.2. Variability in FETs with Multiple SWCNTs

Similar to single SWCNT-FETs, performance variation in FETs with multiple SWCNTs originates from variations in d, Φ_{WF}, L_{ch}, T_{ox}, and defects [87,88,89]. A simple translation of single SWCNT-FET analysis to multiple SWCNT-FET; however, they cannot explain related variations, mostly because of the added complexity coming from variations in the arrangements and types of multiple SWCNTs bridging source/drain contacts. Multiple SWCNT FETs can have random network or aligned SWCNTs with varied density of m- and s-SWCNTs, a wide range of variations in the orientation of SWCNT and its crossings (especially for network SWCNT FET), and a wide distribution in *d*. This section discusses performance variation in aligned array SWCNT-FETs that have significant promise for high performance device applications. Such FETs have negligible variation in SWCNT's orientation [77] and also have negligible SWCNT crossings (Figure 6b). For a discussion on variabilities coming from SWCNT's orientation and crossings, which are more relevant for network SWCNT-FETs, please refer to Ref. [41].

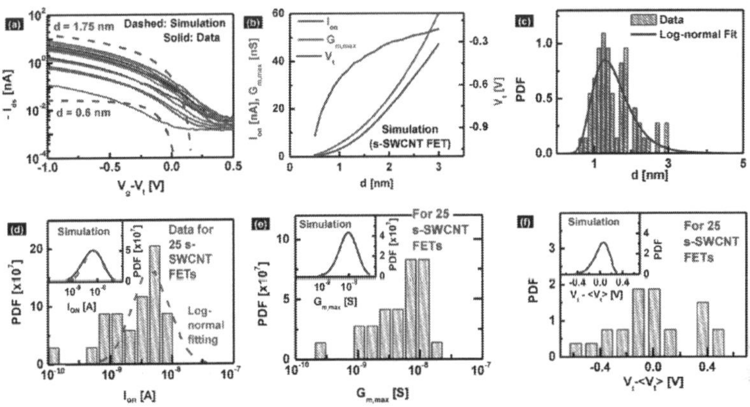

Figure 7. (**a**) Measured I_{ds} *vs.* $V_g - V_t$ ($V_t \equiv V_g$ @ $|I_{ds}| = |I_{ds,max}|/100$) characteristics for single s-SWCNT FETs ($L_{ch} \sim 10$ μm, $V_{ds} = -50$ mV) is within the simulated results for $d = 0.6$ nm and $d = 1.75$nm FETs. (**b**) Simulated I_{on} ($\equiv|I_{ds}|$@ $V_g - V_t = -1$ V, $V_{ds} = -50$ mV), $G_{m,max}$ ($\equiv \max(\partial I_{ds}/\partial V_g)$), and V_t *vs.* *d* for s-SWCNT FETs. (**c**) Diameter distribution that is used to simulate performance distributions of s-SWCNT FETs in (**d–f**). Measured distributions of (**d**) I_{on}, (**e**) $G_{m,max}$, and (**f**) V_t for s-SWCNT FETs agree well with the simulated distributions (insets). For (**f**), $<V_t>$ is the average of V_t distribution. These figures are reprinted with permission from Ref. [44]; © 2012, AIP Publishing LLC.

Solution-processed SWCNTs that are widely used for fabrication of FETs with network and partially aligned SWCNTs have yet to reach the target of 100% chiral selectivity. Some of the resultant FETs, therefore, have high on/off ratios when SWCNT density within the FET is low [35,46]. On the other hand, high performance electronics demand the availability of CVD-grown SWCNTs that always come with a mixture of m- and s-SWCNTs and require the application of post-processing techniques (e.g., electrical breakdown [79], gas-phase reaction [90], thermo-lithography [54]) for having high on/off ratio in multiple SWCNT FETs. In some cases, application of post-processing techniques like electrical breakdown [79] and gas-phase reaction [90] also leads to unwanted removal of s-SWCNTs and, hence, performance degradation or even transistor failure [91]. Considering p_m as the probability of a single SWCNT in the FET being metallic, p_{Rs} as the probability of s-SWCNT being removed via post-processing, *IDC* as the index of dispersion in SWCNT count within the FET, one can simulate the failure probability (p_F) for FETs with aligned SWCNTs at different SWCNT densities (Figure 9a). Such analysis, however, ignores a well-known aspect of 'inferential statistics' [92], *i.e.*, the distributions of SWCNT density and d in macro-scale (population distributions) is not always same as the distributions in micro-scale (sample distributions). As such, performance variability from diameter variations were expected to diminish for high density multiple SWCNT FETs via statistical averaging [25,41,91], making SWCNT density and m-SWCNT's presence as the major contributors to performance variations (Figure 9b) [91].

We performed a comprehensive experimental and theoretical study on FETs with aligned array SWCNTs [44] and ruled out the presence of statistical averaging in aligned-array SWCNT-FETs. Figure 10a plots distribution of I_{on} (measured at a $V_g - V_t = -1$ V) for aligned array-SWCNT FETs having $\langle N \rangle \sim 11$ SWCNTs, where $\langle N \rangle = \langle \rho \rangle W$ is nominal number of SWCNT within the FET, W is the channel width, and $\langle \rho \rangle$ is the average SWCNT density on the substrate per μm across the length of the nanotubes. Since variation in I_{on} follows Poisson statistics [44], we account for changes in μ_{Ion} (that arises from variation in sample preparation) from one set of array-SWCNT FET to another by dividing the standard deviation of $I_{on}/\langle N \rangle$ (σ_{Ion}) with $\sqrt{\mu_{Ion}}$ (where μ_{Ion} is the average of $I_{on}/\langle N \rangle$). Calculated $\sigma_{Ion}/\sqrt{\mu_{Ion}}$, normalized with respect to the same value measured for FETs with

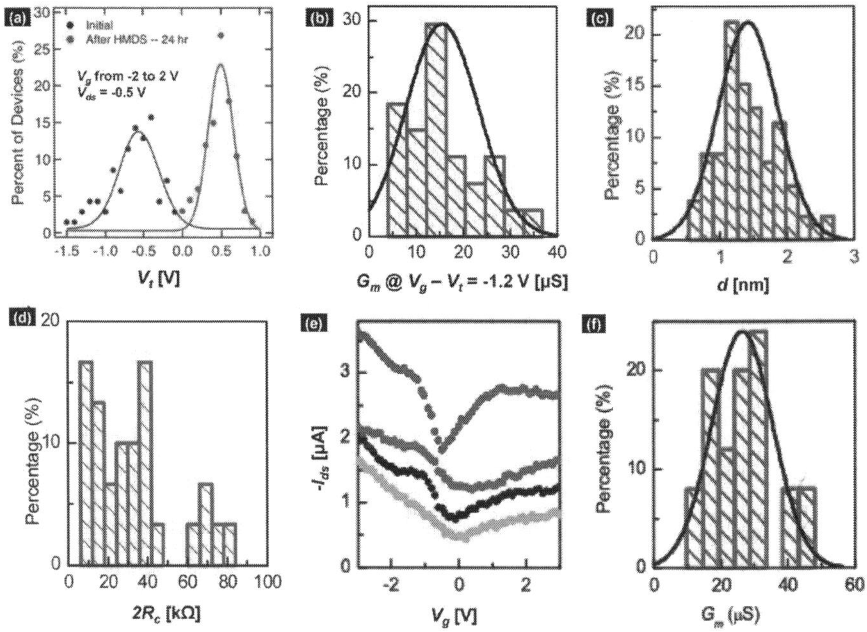

Figure 8. (**a**) Treatment of hexamethyldisilazane (HMDS) for 24 h narrows the V_t distribution significantly. V_g was swept from −2 V to 2 V for p-type, solution-processed s-SWCNT FETs during this measurement. Measured (**b**) G_m distribution and (**c**) diameter distribution for similar solution-processed, s-SWCNT FETs that has small channel length (L_{ch} = 150 nm). (**d**) Extracted contact resistance (R_c) distribution for L_{ch} = 150 nm, solution-processed s-SWCNT FETs. (**e**) Transfer characteristics and (**f**) G_m distribution for solution-processed, m-SWCNT FETs. (**a–c**) and (**d–f**) are reprinted with permission from Refs. [28] and [86], respectively; © 2012 American Chemical Society.

single SWCNT, shows a small reduction with the increase in $<N>$ (Figure 10b). If the diameter and density distributions of SWCNT for each array-SWCNT FETs (sample distribution) were the same as the wafer-level distribution (population distribution) of these parameters, as per central limit theorem [92], the normalized standard deviation should have reduced as $1/\sqrt{<N>}$ due to statistical averaging. The existence of deviation from $1/\sqrt{<N>}$, therefore, suggests significant variations in SWCNTs' density and diameter across the wafer, as was confirmed via extensive atomic force microscopy (AFM) at different locations over a macroscopic area of ST-cut quartz substrate that had CVD-grown aligned arrays of SWCNTs [44].

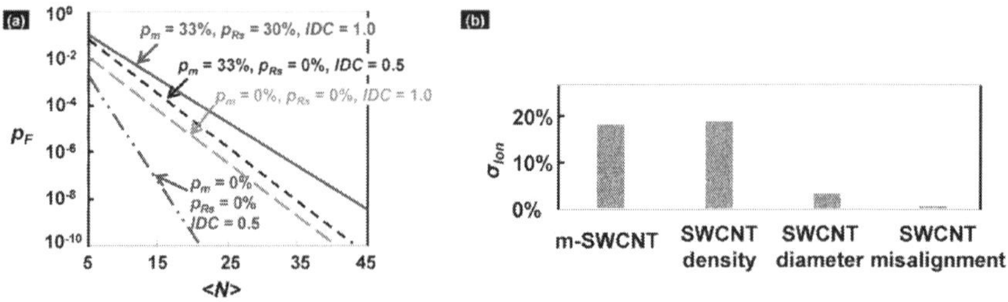

Figure 9. (**a**) Variation in failure probability (p_F; defined as probability of having no s-SWCNT within the FET) of m-SWCNT removed (via post-processing) aligned-array SWCNT-FETs with <*N*> (defined as the average number of SWCNTs within the FET before post-processing). Simulation is performed for different values of p_m (defined as the probability of having m-SWCNT within the FET before post-processing), p_{Rs} (defined as the probability of s-SWCNT removal via post-processing), and *IDC* (index of dispersion in SWCNT count). (**b**) Different variation sources like the presence of m-SWCNT before post-processing, SWCNTs' density, diameter, and misalignment contributes differently to the overall variation in standard deviation of I_{on} (σ_{Ion}). Simulation suggests that the presence of m-SWCNT and density distribution are the key variability sources in SWCNT FETs. These figures are plotted using data from Ref. [91].

To clarify the differences between measured $\sigma_{Ion}/\sqrt{\mu_{Ion}}$ and expected $1/\sqrt{<N>}$ dependency, measured diameter and density variations across the wafer are used first to calibrate the variations measured at the microscopic level (*i.e.*, electronic properties of single-SWNT-FETs, as discussed in section 3.1). Later, the rules of "inferential statistics" [92] is used to analyze variations at the macroscopic level (*i.e.*, for array SWCNT FETs). Here, density variation is eliminated by counting the number of SWCNTs (N) for each array-SWNT FET and then using (I_{ds}/N) – V_g characteristics for standard deviation calculation. Even after accounting for density variation, calculated σ_{Ion} at different <*N*> (normalized to its value for <*N*> = 1) shows significant deviation from $1/\sqrt{<N>}$ scaling (Figure 10c). Such deviation suggests

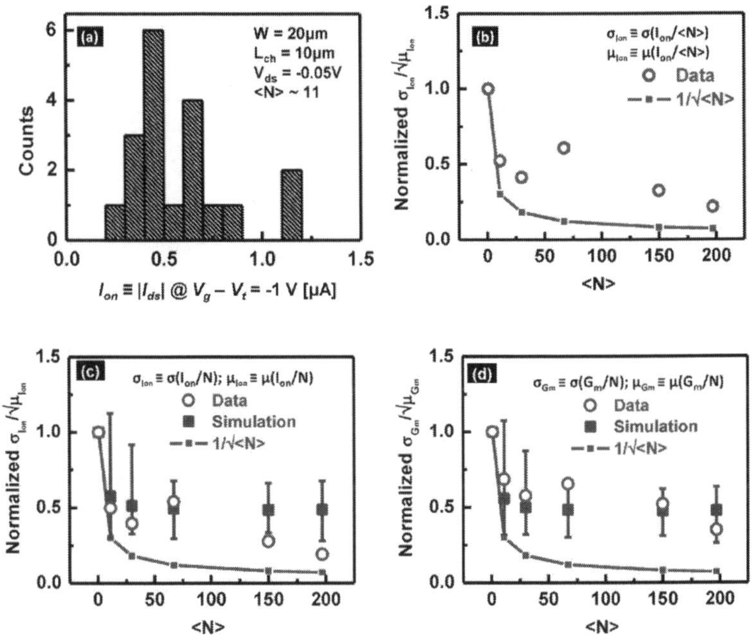

Figure 10. (**a**) Distribution of I_{on} among array-SWCNT FETs with mean SWCNT density <*N*> ~ 11. (**b**) Normalized standard deviation of $I_{on}(\sigma_{Ion})$ *vs.* <*N*> deviates from 1/√<*N*> scaling or central limit theorem. Normalized (**c**) σ_{Ion} and (**d**) σ_{Gm} for array-SWCNT FETs, where the effect of SWCNT density variation is removed by counting *N* for each FET and then calculating the standard deviations using I_{ds}/N–$V_g$characteristics of each FET. Variations in σ_{Ion} and σ_{Gm} still differs from 1/√<*N*> and required the consideration of wafer-scale diameter distribution for matching experimental trends via detailed simulation. Figures are reprinted with permission from Ref. [44]; © 2012, AIP Publishing LLC.

SWCNT density variation as a minor contributor to performance variation, thus contradicting the results of Refs. [25,41,91]. Finally, I_{ds}-$V_g$characteristics of array-SWCNT FETs are simulated for different <*N*> by considering experimentally calibrated diameter dependence of single-SWCNT FETs's I_{ds}-V_g characteristics (Figure 7a) and measured (wafer-scale) density, diameter distributions of SWCNT. Figure 10c,d show normalized standard deviations of I_{on} (σ_{Ion}) and $G_{m,max}$ (σ_{Gm}), as simulated for different <*N*> and their comparison to measured quantities. Since the sample sizes in measuring the standard deviations were small (~17–35), simulated standard deviations show a range of magnitudes for similar sample sizes within which the measured data points fits in. (The simulation framework [44], used for calculating σ_{Ion} and σ_{Gm} of array-SWCNT FETs, neglects contributions from m-SWCNTs. However, the procedure is suitable for demonstrating the importance of wafer-level variations in diameter and density).

Recently, use of smaller equivalent oxide thickness ($EOT \equiv \varepsilon_{SiO2}/\varepsilon_{HK} T_{ox}$; where ε_{SiO2}, ε_{HK} are dielectric constants of SiO_2 and high-κ dielectrics, respectively) has been shown to reduce the performance variation in FETs with single s-SWCNTs [93]. Decreasing *EOT* reduces the width of the Schottky barrier near the SWCNT and source/drain junction [57] and removes the long negative tails in the I_{on} distribution [44]. Moreover, as L_{ch} approaches the carrier mean-free path of SWCNT, I_{ds} saturates for large diameter SWCNTs and hence diameter dependence of I_{ds} is less pronounced [44]. Both these effects are studied by simulating small-scale array-SWCNT-FETs with $EOT \sim 1$ nm and L_{ch} = 300nm. (Here, simulated I_{ds}-V_g [44] is calibrated with measurements of Ref. [63]). Simulation

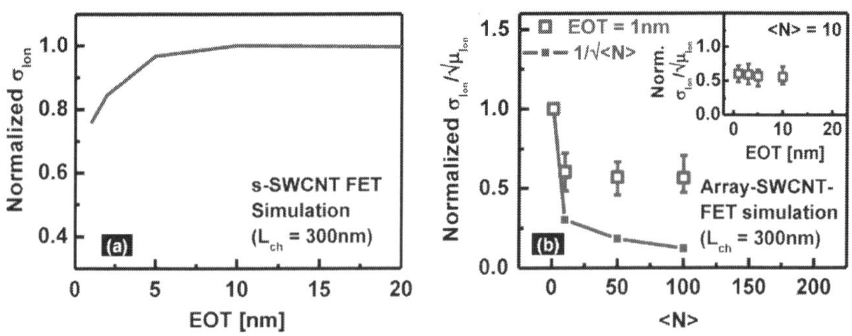

Figure 11. (**a**) Oxide scaling of s-SWCNT FET reduces σ_{Ion} (normalized by its value at equivalent oxide thickness (*EOT* = 20nm), as observed in Ref. [93]. (**b**) Normalized σ_{Ion} (*i.e.*, $\sigma_{Ion}/\sigma_{Ion,<N>=1}$) differs from $1/\sqrt{<N>}$ scaling for *EOT* = 1 nm array-SWNT FETs and show negligible effect of oxide scaling at fixed <*N*> (inset). Figures are reprinted with permission from Ref. [44]; © 2012, AIP Publishing LLC.

suggests a decrease in σ_{Ion} for FETs with <N> = 1 with decreasing *EOT* (Figure 11a). However, at larger <N>, neither *EOT* nor L_{ch} scaling could improve the statistics because the effects of variations in density and diameter remain significant (Figure 11b). Therefore, narrowing diameter and density distributions are identified as the main areas for improvement via advanced growth and/or purification techniques to reduce performance variation in multiple-SWCNT FETs.

4. RELIABILITY OF SWCNT-FETS

Reliability (*i.e.*, device performance degradation) for any device can be interpreted by understanding its origin. For SWCNT-FETs, degradation arises from defects at the interface of SWCNT/oxide and also from defects within the oxide. Trapping and detrapping of carriers (*i.e.*, electrons and holes) into and out of these defects changes performance parameters like V_t, I_{ds}, G_m, *etc.* over time [94,95,96] and also causes hysteresis in the measured I_{ds}-V_g characteristics [83]. In addition to these interface and oxide defects, performance of SWCNT-FET is also affected by topological or Stone-Wales defects [97,98]. They affect the transport of carriers through SWCNT; hence, they lower the mobility of associated FETs. On the other hand, controlled introduction of defects via UV/ozone/chemical treatment [99], doping [20,100], *etc.* enable the application of SWCNTs as sensors by providing low-energy adsorption sites for introduction of chemical species. These structural defects, however, do not evolve with time during the operation of SWCNT-FET, hence do not affect reliability.

In this section, we discuss the nature of interface (section 4.1) and oxide (section 4.2) defects in SWCNT-FETs and show how these defects raise reliability concerns such as hysteresis (section 4.3), low-frequency noise (section 4.4), radiation damage (section 4.5), and time-dependent performance degradation (section 4.6). These interface and oxide defects though cause reliability concerns at different *quantities* depending on the orientation and density of SWCNTs within the FET, the *qualitative* features of reliability remain the same for any orientation and density, as studied in context of hysteresis and performance degradation in Ref. [94]. Therefore, in contrary to Section 3, where variability of FETs with single and multiple SWCNTs is discussed separately, a general discussion of reliability without reference to SWCNTs' density and orientation within the FET is sufficient for this section.

4.1. Interface Defects in SWCNT-FETs

Hydroxyl group (–OH), commonly present at any oxide surface [101,102], is the source of interface defects in SWCNT-FET [83,103,104]. Physisorbed water molecules and oxygen, present at the SWCNT/oxide in ambient condition, control –OH density [102] through electrochemical reaction: $O_2 + 4H^+ + 4e^- \leftrightarrow 2H_2O$ [102,105]. These –OH groups or interface defects surround SWCNTs laying on SiO_2 surface (Figure 12a) and can give rise to large hysteresis in measured I_{ds}-V_g characteristics [83,103,104], as well as potential time-dependent degradation in measured current [95,96]. Application of positive V_{gs} negatively charges –OH groups via electron capture from SWCNT positively shifts V_t and enhances hole conduction in commonly used p-type SWCNT-FET. Application of negative V_g, on the other hand, discharges the –OH group into neutral state, negatively shifts V_t and reduces hole conduction. Marcus-Garischer theory is generally used for modeling charging into or discharging from interface defects in SWCNT-FETs [94,105].

High temperature annealing under vacuum can reduce –OH groups [83,104] by eliminating water molecules at temperatures >190 °C [101]. Annealing at even higher temperature (at >1000 °C [101]) in vacuum would have completely removed the –OH groups; however, they have never been tested because of the high potential for damage to the SWCNTs and to the characterization system at such high temperatures. In addition to annealing in vacuum, –OH group or interface defects can also be reduced by using poly-methyl methacrylate (PMMA) encapsulation with thermal annealing [83] or by growing SWCNTs on different types of hydrophobic surfaces like octadecyltrichlorosilanated-SiO_2 [103,106], hexamethyldisilazanized-SiO_2 [86], octachlorotrisiloxane-capped organic dielectrics [107], *etc.*

4.2. Oxide Defects in SWCNT-FETs

Defects that are present within the gate dielectric, either near the SWCNTs or deep into the dielectric, also affect device performance. These oxide defects are broken bonds (Figure 12b) and pre-exist within the amorphous oxide network at various quantities, depending on the type of oxidation being performed. These pre-existing oxide defects can be annealed, for example, at high temperatures in hydrogen ambient [108]. Oxide defects are also generated near the vicinity of SWCNT during its growth, unless appropriate measures are taken to reduce the defects by running the growth in an inert environment [94]. In addition to the pre-existing ones, oxide defects are also generated during operation of FETs and causes dielectric breakdown [109,110,111]. The nature and origin of these oxide defects have been studied extensively using electron paramagnetic resonance [112,113], stress-induced leakage current [109], sweep-rate dependent I_{ds}-$V_g$hysteresis [94], and optical excitation [114]. Unlike interface defects in SWCNT-FET, oxide defects are ambipolar and can either switch between negative and neutral charged states (*i.e.*, act as electron traps) or switch between positive and neutral charged states (*i.e.*, act as hole traps) [115]—thus, showing a wide range of reliability concerns as discussed below.

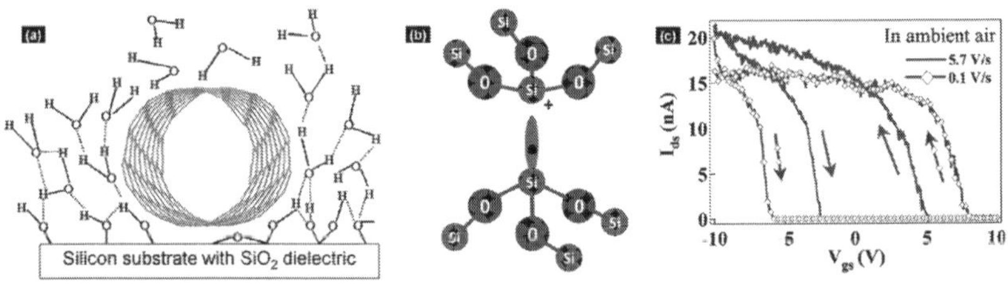

Figure 12. (**a**) Schematic representation of hydroxyl (–OH) group present at the surface of silicon dioxide on which SWCNTs are grown or placed. These –OH groups act as interface defect in SWCNT FETs. (**b**) Schematic representation of oxygen-vacancy related oxide defects [113] within silicon dioxide. (**c**) Transfer characteristics of s-SWCNT FET show large hysteresis between V_g sweeps performed in positive to negative and negative to positive directions. Measured hysteresis depends on sweep rate of V_g. (**a**,**c**) are reprinted with permission from Ref. [83]; © 2003 American Chemical Society.

4.3. Hysteresis in SWCNT-FETs

Large hysteresis in I_{ds}-V_g characteristics is a common phenomenon in SWCNT-FET and its magnitude depends on the concentration of defects in the FET [83,94], density of SWCNT [116], and also on the magnitude [94,107,117] and sweep-rate of V_g [83,94] (Figure 12c).

Two stable charged and discharged states for the defects and switching between these states by application of appropriate V_g is the origin of hysteresis. Though such hysteresis promises memory applications using SWCNT, it is detrimental for device applications.

As mentioned before, both interface and oxide defects can give rise to hysteresis in SWCNT-FET. According to Marcus-Garischer theory [118], –OH groups or interface defects in SWCNT have separate, broad energy levels for discharged (neutral) and negatively charged state, as shown in Figure 13a. In FETs where interface defects predominantly controls hysteresis, the separation between these neutral and charged energy levels of –OH (known as reorganization energy) controls the magnitude of hysteresis [94]. Charging and discharging of oxide defects, on the other hand, can be modeled within a Shockley-Read-Hall framework [119]. Density and type of oxide defects depend on the oxide quality on which carbon nanomaterials are grown or placed on [94]. In some harsh growth

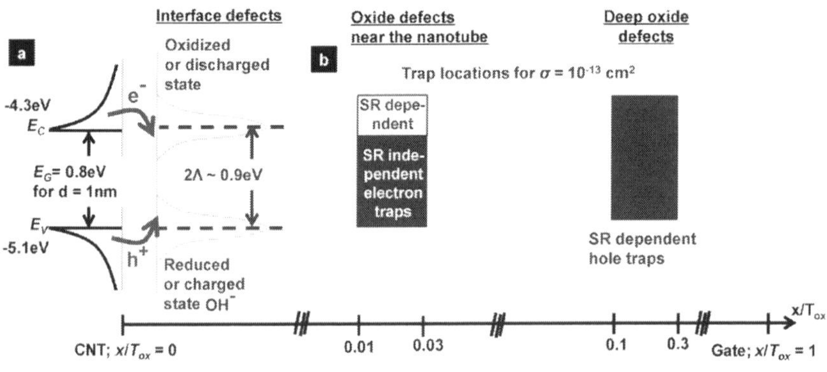

Figure 13. (**a**) Schematic representation of the Marcus-Gerischer theory [118] used to represent the dynamics of –OH group related interface defects to simulate hysteresis in a SWCNT FET with one d = 1 nm s-SWCNT [94]. E_C, E_V, E_G are the conduction band, valence band, bandgap of the s-SWCNT, and Λ is the reorganization energy [105]. (**b**) Representative spatial and energy locations of near-interface electron traps (*i.e.*, oxide defects that switch between negative and neutral charged states) and deep hole traps (*i.e.*, oxide defects that switches between neutral and positive charged states) with σ ~ 10^{-13} cm^2 [120] used for simulating oxide-defect related hysteresis in Ref. [94]; © 2012 WILEY-VCH Verlag GmbH & Co. KGaA, Weinheim.

conditions, oxides near the FET channel are damaged and give rise to large amount of near-interface oxide defects that dominate the trends in measured hysteresis. In other optimized conditions, we only have a small amount of bulk oxide defects that controls the hysteresis trends.

Self-consistent Poisson and drift-diffusion solution [94], supported by detailed experiments, suggest contrasting sweep rate dependence of hysteresis for different types of defects in SWCNT-FET. Interface defects are characterized by short charge capture and emission times compared to the time used for sweeping V_g to measure hysteresis and, therefore, have sweep rate independent hysteresis. On the other hand, near-interface oxide defect have small capture but comparatively larger emission times, such that hysteresis increases with sweep rate; whereas oxide defects within the bulk of the dielectric have large capture and emission times, such that hysteresis decreases with sweep rate. Measurement of hysteresis at different sweep-rate dependencies, therefore, is used to identify the contributions from different types of defects on hysteresis in SWCNT-FET and modify the fabrication steps appropriately to optimize different defects. Growing SWCNTs in an inert environment reduces the amount of near-interface oxide defects [94] and enables one to employ different interface defect removal techniques to compare their effectiveness. A comparison of different –OH group reduction methods (e.g., heating FETs in vacuum [83], coating SWCNT with PMMA [83] or spin-on-glass [94]; Figure 14a) suggests high temperature annealing, quality of coatings to block water penetration, and reaction of OH-group with coated molecules as the key for reducing hysteresis originating from –OH group related interface defects. All these –OH removal techniques have shown negligible hysteresis when SWCNT-FET is swept with smaller $|V_{g,max}|$. Table 1 summarizes the values of $|V_{g,max}|$ and corresponding maximum of average oxide electric fields ($E_{ox,max}$) that has been reported to show negligible hysteresis under various conditions [9,83,94,103,106,107,121,122]. ($E_{ox,max}$calculation steps for all the references are shown in the caption of Table 1.) Among these conditions, SWCNT-FETs with spin-on-glass/HfO_2 gate dielectric shows negligible hysteresis for highest $E_{ox,max} \sim$ 0.45 MV/cm. At higher $|V_{g,max}|$ or $E_{ox,max}$, hysteresis reappears [94,107] which suggests incomplete removal of interface and oxide defects using the techniques mentioned above. Use of combinatorial approaches such as placing SWCNT on hydrophobic surfaces, annealing at temperatures above 400 °C in an inert environment, and then coating with hydrophobic gate dielectric may eliminate hysteresis even at higher $E_{ox,max}$. Hysteresis reduction is critical for SWCNT-FET, because reduced hysteresis is generally accompanied with a narrower distribution in V_t (Figure 8a) and also for the hysteresis itself (Figure 14b) [86]. Removal of –OH group from SWCNT-FETs, therefore, is necessary to improve variability in SWCNT-FET.

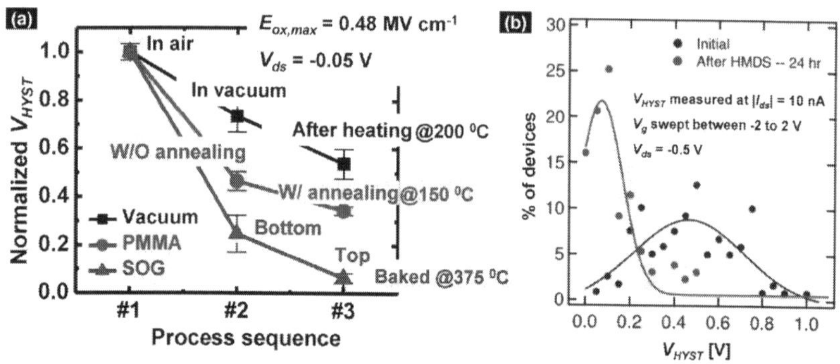

Figure 14. (**a**) Comparison of effectiveness of different hysteresis reduction techniques like heating in vacuum, use of poly-methyl methacrylate (PMMA) and SOG. Measured hysteresis (V_{HYST}) in each case at similar $E_{ox,max}$ is normalized with that measured in air. For heating in vacuum, hysteresis is first measured in air at room temperature, then after 10 h in vacuum (10^{-5} Torr), and finally after 10 h of heating at 200 °C followed by cooling to room temperature in vacuum. For PMMA passivation (with: W/ and without: W/O), hysteresis is first measured in air before PMMA coating, then after PMMA coating and heating for 1min at 110 °C, and finally after an additional 15 h annealing at 150°C. For SOG encapsulation, hysteresis is first measured in air using a bottom-gated configuration, then after SOG encapsulation and 375 °C baking using bottom- and top-gated configurations. This figure is taken from Ref. [94]; © 2012 WILEY-VCH Verlag GmbH & Co. KGaA, Weinheim. (**b**) Use of hysteresis reduction technique like HMDS treatment before placing SWCNTs on oxides is also effective for narrowing distribution of measured hysteresis. This figure is reprinted with permission from Ref. [86]; © 2012 American Chemical Society.

Table 1. Values of EOT, SWCNT density (D), capacitance ratio (Ξ) [131], maximum electric field with parallel plate assumption (*i.e.*, SWCNT is presumed to form a dense film), and maximum of the average oxide electric field ($E_{ox,max}$) for reported instances [9,83,94,103,106,107,121,122] of negligible hysteresis in SWCNT-FETs. Here, $\Xi = 4\pi \times EOT/\varepsilon_{SiO2} \times \lambda_0[2/\varepsilon_{SiO2} \times ln[\lambda_0/d \times sinh(2\pi \times EOT/\lambda_0)/\pi] + C_Q^{-1}]$ [131] (where $\lambda_0 = 1000/D$ is the average distance between each SWCNT, $C_Q = 4 \times 10^{-12}$ F/cm is the quantum capacitance [132]), $E_{P,max} = V_{G,max}/EOT$, and $E_{ox,max} = \Xi \times E_{P,max}$.

Reference	EOT [nm]	D [μm^{-1}]	Ξ	$V_{g,max}$ [V]	$E_{P,max}$ [MV cm^{-1}]	$E_{ox,max}$ [MV cm^{-1}]	Type of Dielectric
[9] Cao *et al.*, Nature '08	8.2	~6	0.01	2	2.44	0.02	HfO_2
[106] Weitz *et al.*, NL '07	20.3	5.48	0.16	2	0.98	0.16	Octadecyltrichlorosilanated-SiO_2
[107] Hur *et al.*, JACS '05	6.9	1 (Single CNT)	0.01	1	1.45	0.02	Octachlorotrisiloxane-capped organic dielectric
[83] Kim *et al.*, NL'03	500	1 (Single CNT)	0.36	10	0.2	0.07	SiO_2 with PMMA coating
[103] McGill *et al.*, APL '06	100	~3	0.30	2	0.2	0.06	Octadecyltrichlorosilanated-SiO_2
[121] Hu *et al.*, Phys. E '08	200	~3	0.45	10	0.5	0.23	Aminopropyltriethoxysilaned-SiO_2
[131] Kim *et al.*, APL '07	6.5	1 (Single CNT)	0.01	1	1.54	0.02	Al_2O_3
[94] Jin *et al.*, AFM '12	40	~6	0.29	6	1.5	0.45	Spin-on-glass (SOG) + HfO_2

4.4. Low-Frequency Noise in SWCNT-FETs

Low-frequency noise measurement is a general way of characterizing deformations or defects in any nanoscale systems [123,124,125,126,127]. It is also a reliability concern for flash memories [128,129], radio-frequency (RF) circuits [130], and, in general, for many different digital and analog applications employing nanoscale FETs. Among different components of low-frequency noise, the following components are routinely measured: (i) the frequency independent thermal noise component with power spectral density of $4k_BTR$, which originates from the statistical Brownian motion of carriers at temperature T in a resistive system with resistance R; (ii) the Lorenzian-type noise component with power spectral density ~ $1/[1 + (2\pi f\tau)^2]$, which originates from trapping/detrapping events for a small number of deformations or defects with time constant τ and cut-off frequency $f_c = 1/\tau$; and (iii) the flicker noise component with power spectral density ~ $1/f^n$, which originates from a distribution of trapping/detrapping events.

Noise in SWCNT-FETs has been widely studied to understand the origin of deformations or defects in SWCNT lattice and in its surrounding oxides [123,124,125,128,129], and also to explore the effect of radiation damage [133] and chemical exposure [134]. Measured low-frequency noise in SWCNT is much more pronounced than in silicon electronics [135,136] because of the presence of a few carriers for conduction [134]. In addition to structural defects in the SWCNT lattice [135], oxide defects contribute significantly to low-frequency noise behaviors in SWCNT-FETs [137]. (A recent discussion on different sources of low-frequency noise in SWCNT materials and devices can be found in Ref. [138].) Interface defects interact at faster time-scales and hence noise because interface defects are generally overwhelmed by thermal noise at high frequencies. Oxide defects, on the other hand, have comparatively longer and distributed charging and discharging times (either due to distribution in tunneling distance [126,127] or due to distribution in defects' energetics [139]) that give rise to $1/f$ behavior in the power spectral density of measured noise in I_{ds} at low frequencies [134,135], as shown in Figure 15a, which is measured for a single s-SWCNT-FET at different V_g[134]. From Hooge's number fluctuation theory [140] noise power spectral density (S_I) can be expressed as,

$$S_I = I_{ds}^{\ 2}\frac{A}{f}, \text{ where } A \sim \frac{\alpha_H}{I_{ds}}. \tag{3}$$

where α_H is a fitting parameter that varies from device to device. Equation (3) suggests a reduction in noise amplitude A with the increase in carrier number or I_{ds}, as measured for the p-type FET for different V_g(Figure 15b) [134].

Noise measurements in cryogenic conditions reduce thermal noise and also freeze-off some of the defects that contribute to noise at higher temperatures. Responses from the remaining few oxide defects result in random telegraph noise (RTN) with stochastic variation in I_{ds} and discrete current levels [137,141,142,143].Figure 16a shows RTN obtained from a SWNT-FET (measured at T = 4.2 °K) suggesting four major switching levels resulting from a slow switching defect (whose fluctuation is digitized in the bottom of Figure 16a) and a fast

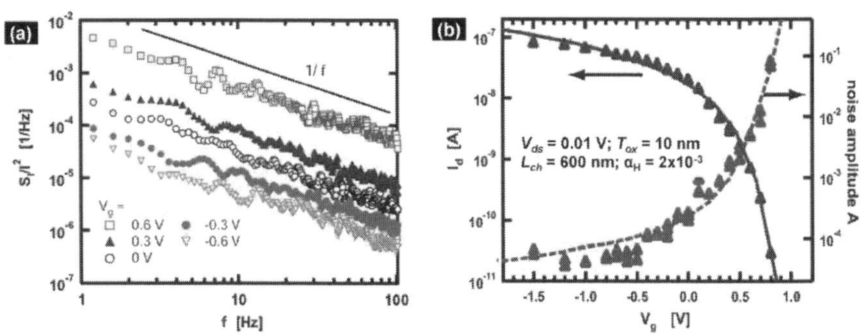

Figure 15. (**a**) Flicker or $1/f$ noise measured for single s-SWCNT FET at different V_g. (**b**) Increase in V_g decreases I_{ds} and hence increases noise magnitude following Hooge's number fluctuation theory [140]. These figures are reprinted with permission from Ref. [134]; © 2006 American Chemical Society.

switching defect [142]. Corresponding power density spectrum for these fluctuations is shown inFigure 16b that suggests two overlapping Lorenzian spectrums with corner frequencies (f_c) at 0.2 Hz (for the slow defect) and 5.8 Hz (for the fast defect). At higher temperatures (T = 60 °K), however, responses from these two defects are mixed up (inset of Figure 16b) and noise from new defects (with higher f_c) shows up. Similar noise analysis at different V_g (Figure 16c) enables one to extract distributions of capture time τ_c(defined as the time required for a high to low transition in I_{ds}) and emission time τ_e (defined as the time required for a low to high transition in I_{ds}). These distributions are used to calculate the averages of capture and emission times ($<\tau_c>$ and $<\tau_e>$) at different V_g (Figure 16d) and, hence, the location of defect causing RTN from the slope of $\ln(<\tau_e>/<\tau_c>)$ *vs.* V_g plot [137].

Though RTN analysis has not yet been employed to study the evolution of defects—*i.e.*, reliability in SWCNT-FETs—$1/f$ noise has been used to study the radiation effects in such FETs [133]. An increase in $1/f$noise after gamma radiation (Figure 17) suggests generation of defects within the oxide. The next section discusses the radiation effects on SWCNT-FETs in detail.

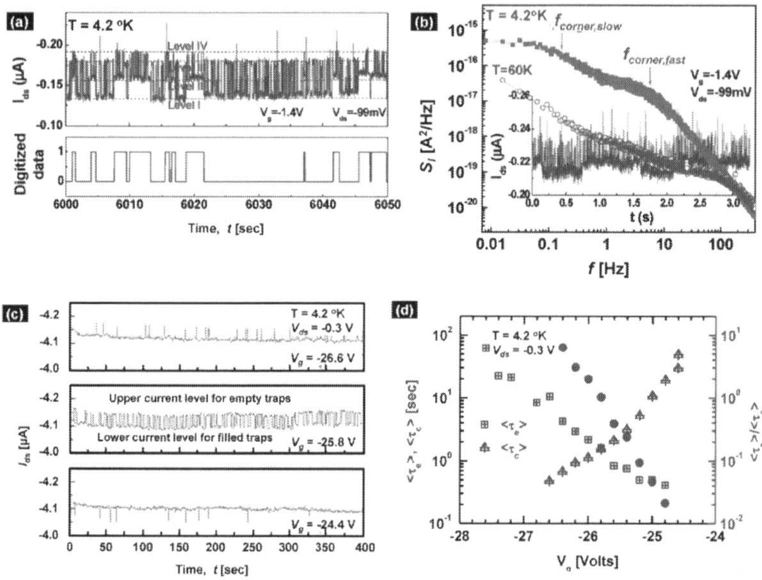

Figure 16. (**a**) Random telegraph noise (RTN) measured in a SWCNT-FET at 4.2 °K with V_g = −1.4 V and V_{ds} = −99 mV shows four major switching levels (red dashed lines) due to the presence of one slow (with larger trapping/detrapping time) and one fast (with smaller trapping/detrapping time) defect. The digitized fluctuation for the slow defect is shown in the bottom graph. (**b**) Noise power spectral density ofFigure 16a at 4.2 °K and also for the same device at 60 °K. The inset shows the time domain current data at 60 °K. Spectral density at 4.2 °K shows the cut-off frequencies designating the presence of slow and fast traps. Response from these defects disappears at higher temperatures. (**c**) Low temperature RTN measurements in another SWCNT-FET at different V_g. (**d**) Voltage dependence of average capture $<\tau_c>$, emission $<\tau_e>$ times and their ratios. (**a**,**b**) are reprinted with permission from Ref. [142]; © 2008 American Chemical Society and (**c**,**d**) are replotted using data from Ref. [137].

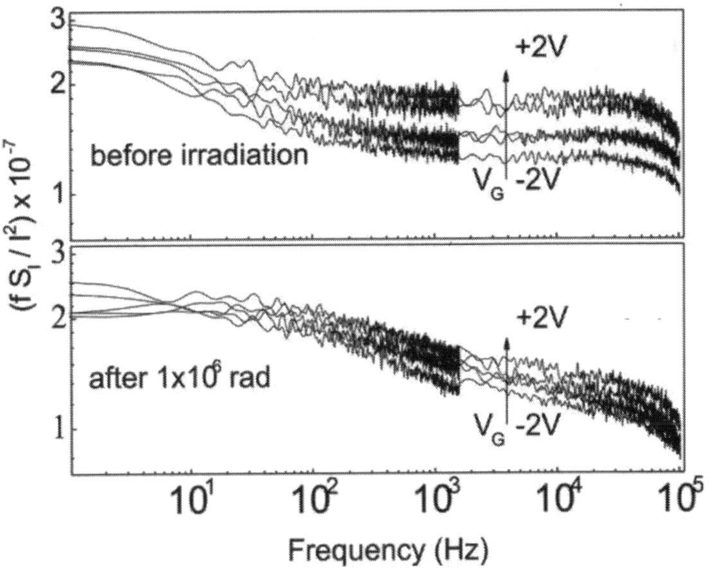

Figure 17. Measured noise power spectral density at different V_g before and after gamma radiation (reprinted with permission from Ref. [133]; © 2010, AIP Publishing LLC).

4.5. Radiation Damage in SWCNT-FETs

Promising performance parameters of SWCNTs have initiated studies to explore their potential in terrestrial and space-based applications [133,144,145,146,147,148,149]. In such niche applications, radiation tolerance is one of the key reliability parameters. In this regard, use of SWCNT as a channel showed significant advantages over its silicon-based predecessors. Small cross-sections of SWCNT make it less susceptible to incident radiation particles, unless radiation dose (termed as displacement damage dose, DDD) exceeds ~ 10^{13} MeV/g [150]. One can evaluate DDD for different types of particles (e.g., alpha, proton, ionized chemical species) by considering the number of particles striking SWCNTs per unit area (*i.e.*, fluence) and the efficiency of these particles in creating defects within the SWCNT lattice (which is quantified using the term non-ionizing energy loss [150]). Damage in SWCNT at high radiation dose (with DDD > 10^{13} MeV/g) can be monitored using Raman spectroscopy as an increase in intensity at the defect (D)-band wavenumber of ~ 1300 cm^{-1} [133,146,147,148,151] and also using electrical measurement as an increase in resistance [148,151]. Figure 18a,b show Raman spectra for m- and s-SWCNT-FETs measured at different amount of ionized boron irradiation. Radiation gradually introduces defects within the SWCNT lattice, therefore, increasing the ratio of Raman intensities measured for D- and G^+-bands or for D- and G'-bands. (G^+- Raman signal comes from the atomic vibration of carbon atoms along the axial direction of SWCNT and peaks at a wavenumber of ~1590 cm^{-1} with no diameter dependence. On the other hand, G'- Raman signal arises from a two-photon, inter-valley, second-order Raman scattering process and peaks at a wavenumber of ~2700 cm^{-1} with significant diameter dependence. Both G^+ and G' have no influence from structural defects in SWCNTs. (Dresselhaus *et. al* [152] provide an elaborate review of Raman signal and its origin in different types of SWCNTs.) Ratio of Raman intensities obtained using D- and G'-bands (D/G') are shown to be more sensitive in monitoring the effect of radiation that linearly depends on DDD (Figure 18c), when D/G' at a particular amount of radiation (Φ) is normalized to its value measured before radiation (Φ = 0) [148]. The same linear degradation law also holds for the radiation-induced resistance degradation in SWCNT and graphene.

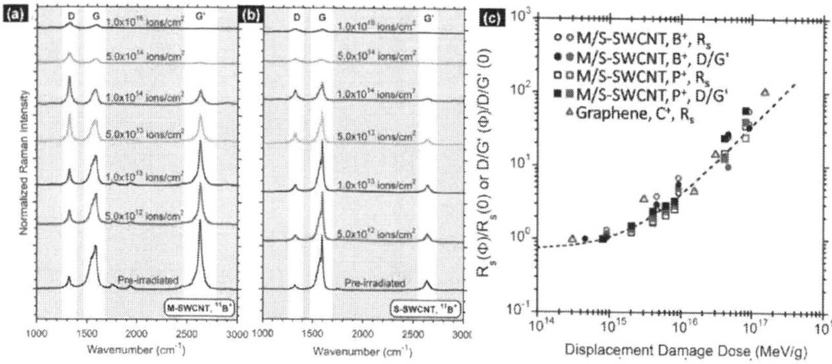

Figure 18. Raman spectra for (**a**) m-SWCNT and (**b**) s-SWCNT thin-films irradiated with 150 keV $^{11}B^+$ ion. The pre-irradiated spectrum is given as a reference and the intensity of each irradiated sample is normalized to the G band intensity of the reference. (**c**) Changes in the normalized (with respect to pre-radiated sample) Raman D/G′ ratio and normalized sheet resistance (R_s) with displacement damage dose (DDD) for boron-ion (B^+) and phosphorous ion (P^+) irradiated m-SWCNTs and s-SWCNTs, and carbon-ion (C^+) irradiated graphene (Ref. [153]). The dashed line is a linear fit of all D/G' ratioand R_s values. Figures are reprinted with permission from Ref. [148]; © 2012, AIP Publishing LLC.

At radiation doses comparable to aerospace environments (*i.e.*, for DDD<10^{13} MeV/g), only the dielectric layer of SWCNT-FETs are damaged with no damage to the SWCNT lattice [133,146,154]. Radiation in such condition changes V_t of SWCNT-FETs with negligible change in mobility (Figure 19a), therefore, suggests negligible defect formation within the SWCNT. The behavior is also confirmed via Raman spectroscopy (Figure 19b) that shows negligible change in D-band and the radial breathing mode (RBM) Raman signals. (RBM signal comes from the atomic vibration of carbon atoms along the radial of SWCNT and has peaks at wavenumbers in the range of ~100–400 cm^{-1} with strong diameter dependence [152].) A decrease in V_t with increasing radiation for the p-type SWCNT-FET (Figure 19a) suggests creation of oxide defects that act as hole trapping sites. The situation improves significantly when a comparatively thinner (~23 nm) silicon oxynitride dielectric (grown via plasma-enhanced chemical vapor deposition) replaces the regularly used ~100 nm SiO_2 in making SWCNT-FETs for radiation testing [145]. Pre- and post-radiation I_{ds}-$V_g$measurements (Figure 20) suggest negligible change in FET parameters above threshold (*i.e.*, V_t and μ_{eff}). Creation of balanced hole and electron trapping sites within the dielectric [155] and use of thin dielectric that allows the escape of trapped charges [156] are presumed to be the origin of such

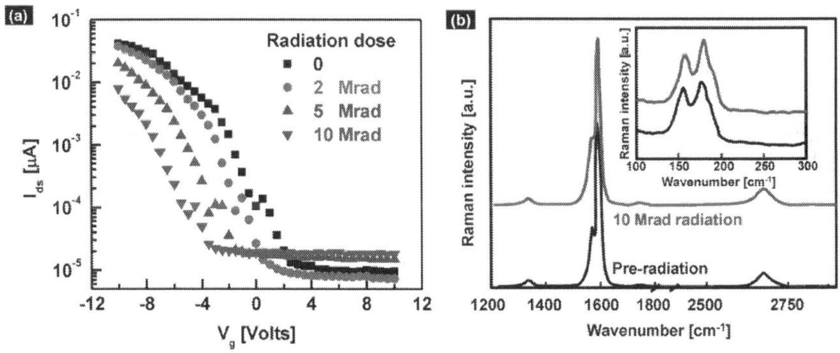

Figure 19. (**a**) Transfer characteristics of an SWCNT-FET measured in vacuum for different doses of Co-60 gamma rays. (**b**) G-band and G'-band Raman spectra of SWCNTs (98% s-SWCNT enriched thin film) within the FET before (pre-) and after (post-) irradiation in vacuum. Inset shows the Raman spectra in the low wavenumber regime that suggests negligible change in radial breathing modes. These figures are replotted using data from Ref. [146].

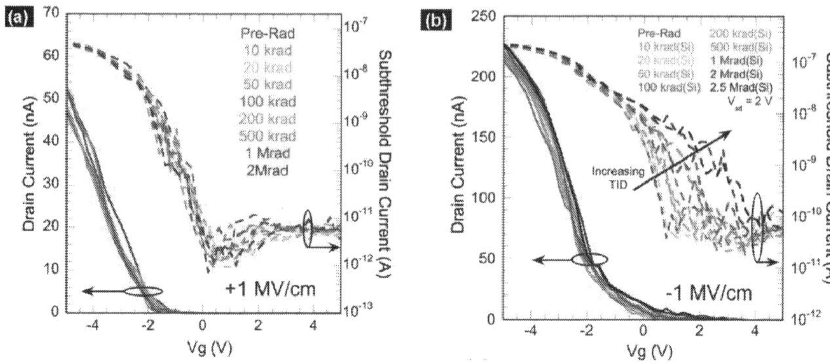

Figure 20. Transfer characteristics of an s-SWCNT FET fabricated with radiation-hard silicon oxynitride gate dielectric. Irradiation is conducted under vacuum with **(a)** positive, and **(b)** negative gate electric field of 1 MV/cm. Figures are reprinted with permission from Ref. [145] © Cambridge University Press 2011.

radiation hardness in these devices. The sub-threshold behavior, on the other hand, shows degradation when the devices are exposed to a negative gate electric field of −1 MV/cm. The origin of such degradation has been attributed to the formation of electron trapping sites within the oxide or at the interface of SWCNT—which requires detailed future analysis. Overall, SWCNT-FETs with thin dielectric offer great potential for space electronics in the future.

4.6. Performance Degradation in SWCNT-FETs

Performance degradation of transistor parameters over time is a critical reliability concern that dictates the determination of device lifetime. In transistor reliability literature, bias temperature instability [157] and stress-induced leakage current [109] are often used to monitor performance degradation and to probe the dynamics of defects in different regions of the FET. Though hysteresis and low-frequency noise has been widely used to study the nature of defects in SWCNT-FET, detailed analysis of performance degradation over longer time scales is still lacking. Studies of degradation up to a few seconds [95,96] though provides critical information about the nature of some of the defects, projection of performance degradation up to an intended lifetime of 5–10 years requires accelerated tests (at voltages much higher than use condition) and for longer times (1000 s or more), which are widely used in conventional CMOS reliability studies [157].

Some of the recent long-term performance degradation studies in SWCNT-FETs [94] suggest oxide defect as the key contributor for the instability. SWCNT-FETs with a large amount of near-interface oxide defects show significant I_{ds} degradation (ΔI_{ds}; Figure 21a) and have a small time constant $\tau \sim 7 \times 10^4$ sec for the fitted stretched exponential forms according to $\Delta I_{ds} = I_{ds0}[1 - \exp(-t/\tau)^{\beta}]$ (where I_{ds0} is the magnitude of drain current at $t = 0$, t is the time at which degradation is measured, and $\tau, \beta \sim 0.35$ are fitting parameters) [158]. On the other hand, devices with a small amount of bulk oxide defects have lower ΔI_{ds} and long $\tau \sim 2 \times 10^7$ s (with $\beta \sim 0.35$). Moreover, the nature of trapping into oxide defects remains the same at different accelerated stress conditions, as seen from the measured ΔI_{ds} at different gate biases (Figure 21b). All these degradation curves can be fitted with a fixed β and a voltage dependent τ (higher for higher V_g) that suggests more and deeper oxide defects are accessible at higher stress biases. Therefore, by using τ as a voltage dependent parameter, one can project reliability of these SWCNT-FETs at different use conditions. Detailed analysis of such degradation by understanding the trapping dynamics (which may require the use of one or multiple time constants [95]) is still lacking.

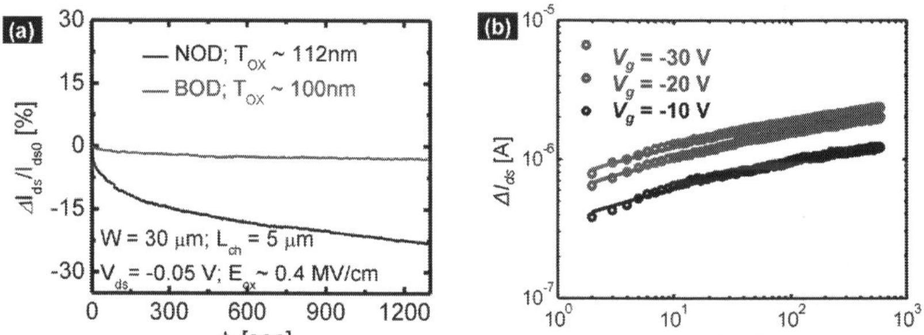

Figure 21. (**a**) Percentage change in I_{ds} (ΔI_{ds}) relative to its initial value (I_{ds0}) as a function of time (t) for two SWCNT-FETs, one having a large amount of near-interface oxide defects (NOD) and the other having a small amount of bulk oxide defects (BOD). The figure is taken from Ref. [94]; © 2012 WILEY-VCH Verlag GmbH & Co. KGaA, Weinheim. (**b**) Drain current degradation measured at different gate stress bias V_g. All the degradation curves can be fitted (lines) using a single expression: $\Delta I_{ds} = I_{ds0}[1 - \exp(-t/\tau)^{\beta}]$ with fixed β and voltage dependent τ (higher for higher V_g).

5. SUMMARY AND OUTLOOK

We have discussed different sources of variability and reliability that controls the change in time-zero and time-dependent performance parameters in SWCNT-FETs. Variability in SWCNT FETs performance is mainly attributed to the distributions of SWCNTs' diameter across the wafer and the presence of hydroxyl-group related interface defects in SWCNTs' surroundings. The removal of m-SWCNTs and the tightening of diameter distribution for s-SWCNTs via chiral-selective growth [159] or post-processing [50,54], and complete passivation of SWCNTs via removal of hydroxyl groups [83,94] are necessary for controlling variability. Removal of hydroxyl groups from SWCNTs' surfaces will also be the key to having a reliable SWCNT FET that will operate with reduced performance degradation over its lifetime. In addition, oxide defects keep playing a dominant role in controlling reliability parameters, like low-frequency noise, radiation damage, and long-term performance degradation. Therefore, solutions in terms of having reliable gate dielectrics with less defects [111,160] will not only help current CMOS scaling, but will also assist with the introduction of beyond CMOS devices with SWCNT-based FET.

ACKNOWLEDGMENTS

The author acknowledges contributions from John Rogers (University of Illinois) for introducing the presented area of research and also from Sunghun Jin, Frank Du (University of Illinois) for related experiments and discussion. Discussions with Muhammad A. Alam (Purdue University), Cory D. Cress (Naval Research Laboratory), Benji Maruyama (Air Force Research Laboratory), and funding from Air Force Office of Scientific Research (contract number: FA 8650-09-D-5037), Materials Structures and Devices Focus Center (Task ID: 2051.003.4) of Semiconductor Research Corporation are acknowledged for preparation of this review article.

CONFLICTS OF INTEREST

The authors declare no conflict of interest.

REFERENCES

1. Iijima, S. Helical microtubules of graphitic carbon. Nature 1991, 354, 56–58.

2. Datta, S. Quantum Transport: Atom to Transistor, 2nd ed.; Cambridge University Press: Cambridge CB2 2RU, UK, 2005.

3. Saito, R.; Sato, K.; Oyama, Y.; Jiang, J.; Samsonidze, G.G.; Dresselhaus, G.; Dresselhaus, M.S. Cutting lines near the Fermi energy of single-wall carbon nanotubes. Phys. Rev. B 2005, 72.

4. Jorio, A.; Pimenta, M.A.; Souza, A.G.; Saito, R.; Dresselhaus, G.; Dresselhaus, M.S. Characterizing carbon nanotube samples with resonance Raman scattering. New J. Phys. 2003, 5.

5. Zhou, X.J.; Park, J.Y.; Huang, S.M.; Liu, J.; McEuen, P.L. Band structure, phonon scattering, and the performance limit of single-walled carbon nanotube transistors. Phys. Rev. Lett. 2005, 95.

6. Durkop, T.; Getty, S.A.; Cobas, E.; Fuhrer, M.S. Extraordinary mobility in semiconducting carbon nanotubes. Nano Lett. 2004, 4, 35–39.

7. Guo, J.; Goasguen, S.; Lundstrom, M.; Datta, S. Metal-insulator-semiconductor electrostatics of carbon nanotubes. Appl. Phys. Lett. 2002, 81, 1486–1488.

8. Rogers, J.A.; Someya, T.; Huang, Y. Materials and mechanics for stretchable electronics. Science 2010,327, 1603–1607.

9. Cao, Q.; Kim, H.S.; Pimparkar, N.; Kulkarni, J.P.; Wang, C.J.; Shim, M.; Roy, K.; Alam, M.A.; Rogers, J.A. Medium-scale carbon nanotube thin-film integrated circuits on flexible plastic substrates. Nature2008, 454, 495–500.

10. Sun, Y.; Rogers, J.A. Inorganic semiconductors for flexible electronics. Adv. Mater. 2007, 19, 897–1916.

11. Perelaer, J.; Smith, P.J.; Mager, D.; Soltman, D.; Volkman, S.K.; Subramanian, V.; Korvink, J.G.; Schubert, U.S. Printed electronics: The challenges involved in printing devices, interconnects, and contacts based on inorganic materials. J. Mater.Chem. 2010, 20, 8446–8453.

12. Cao, Q.; Hur, S.H.; Zhu, Z.T.; Sun, Y.G.; Wang, C.J.; Meitl, M.A.; Shim, M.; Rogers, J.A. Highly bendable, transparent thin-film transistors that use carbon-nanotube-based conductors and semiconductors with elastomeric dielectrics. Adv. Mater. 2006, 18, 304–309.

13. Xiong, F.; Liao, A.D.; Estrada, D.; Pop, E. Low-power switching of phase-change materials with carbon nanotube electrodes. Science 2011, 332, 568–570.

14. Kianian, S.; Rosendale, G.; Manning, M.; Hamilton, D.; Huang, X.M.H.; Robinson, K.; Kim, Y.W.; Rueckes, T. A 3D stackable Carbon Nanotube-based nonvolatile memory (NRAM). In Proceedings of the ESSDERC, Sevilla, Spain, 14–16 September, 2010; pp. 404–407.

15. Rosendale, G.; Kianian, S.; Manning, M.; Hamilton, D.; Huang, X.M.H.; Robinson, K.; Kim, Y.W.; Rueckes, T. A 4 Megabit Carbon Nanotube-based nonvolatile memory (NRAM). In Proceedings of the ESSDERC, Sevilla, Spain, 14–16 September, 2010; pp. 478–481.

16. Fuhrer, M.S.; Kim, B.M.; Durkop, T.; Brintlinger, T. High-mobility nanotube transistor memory. Nano Lett. 2002, 2, 755–759.

17. Di Bartolomeo, A.; Yang, Y.; Rinzan, M.B.M.; Boyd, A.K.; Barbara, P. Record endurance for single-walled carbon nanotube-based memory cell. Nanoscale Res. Lett. 2010, 5, 1852–1855.

18. Jarosz, P.R.; Shaukat, A.; Schauerman, C.M.; Cress, C.D.; Kladitis, P.E.; Ridgley, R.D.; Landi, B.J. High-performance, lightweight coaxial cable from carbon nanotube conductors. ACS Appl. Mater. Interfaces2012, 4, 1103–1109.

19. Nair, P.R.; Alam, M.A. Performance limits of nanobiosensors. Appl. Phys. Lett. 2006, 88, 233120:1–233120:3.

20. Pengfei, Q.F.; Vermesh, O.; Grecu, M.; Javey, A.; Wang, O.; Dai, H.J.; Peng, S.; Cho, K.J. Toward large arrays of multiplex functionalized carbon nanotube sensors for highly sensitive and selective molecular detection. Nano Lett. 2003, 3, 347–351.

21. Collins, P.G.; Bradley, K.; Ishigami, M.; Zettl, A. Extreme oxygen sensitivity of electronic properties of carbon nanotubes. Science 2000, 287, 1801–1804.

22. Alam, M. Reliability- and process-variation aware design of integrated circuits. Microelectron. Reliab.2008, 48, 1114–1122.

23. Lee, Y.H.; McMahon, W.; Mielke, N.; Lu, Y.L.R.; Walstra, S. Managing bias-temperature instability for product reliability. Proc. Int. Symp. VLSI Technol. 2007, 1, 52–53.

24. Moore, G.E. Cramming more components onto integrated circuits (Reprinted from Electronics, pg 114–117, April 19, 1965). Proc. IEEE 1998, 86, 82–85.

25. Kang, S.J.; Kocabas, C.; Ozel, T.; Shim, M.; Pimparkar, N.; Alam, M.A.; Rotkin, S.V.; Rogers, J.A. High-performance electronics using dense, perfectly aligned arrays of single-walled carbon nanotubes. Nat. Nanotechnol. 2007, 2, 230–236.

26. Patil, N.; Lin, A.; Myers, E.R.; Ryu, K.; Badmaev, A.; Zhou, C.W.; Wong, H.S.P.; Mitra, S. Wafer-scale growth and transfer of aligned single-walled carbon nanotubes. IEEE Trans. Nanotechnol. 2009, 8, 498–504.

27. Wang, C.; Badmaev, A.; Jooyaie, A.; Bao, M.Q.; Wang, K.L.; Galatsis, K.; Zhou, C.W. Radio frequency and linearity performance of transistors using high-purity semiconducting carbon nanotubes. ACS Nano 2011, 5, 4169–4176.

28. Cao, Q.; Han, S.-J.; Tulevski, G.S.; Franklin, A.D.; Haensch, W. Evaluation of field-effect mobility and contact resistance of transistors that use solution-processed single-walled carbon nanotubes. ACS Nano 2012, 6, 6471–6477.

29. Vijayaraghavan, A.; Blatt, S.; Weissenberger, D.; Oron-Carl, M.; Hennrich, F.; Gerthsen, D.; Hahn, H.; Krupke, R. Ultra-large-scale directed assembly of single-walled carbon nanotube devices. Nano Lett.2007, 7, 1556–1560.

30. Bardecker, J.A.; Afzali, A.; Tulevski, G.S.; Graham, T.; Hannon, J.B.; Jen, A.K.Y. Directed assembly of single-walled carbon nanotubes via drop-casting onto a UV-patterned photosensitive monolayer. J. Am. Chem. Soc. 2008, 130, 7226–7227.

31. Park, H.; Afzali, A.; Han, S.-J.; Tulevski, G.S.; Franklin, A.D.; Tersoff, J.; Hannon, J.B.; Haensch, W. High-density integration of carbon nanotubes via chemical self-assembly. Nat. Nanotechnol. 2012, 7, 787–791.

32. Cao, J.; Nyffeler, C.; Lister, K.; Ionescu, A.M. Resist-assisted assembly of single-walled carbon nanotube devices with nanoscale precision. Carbon 2012, 50, 1720–1726.

33. Wang, J. Carbon-nanotube based electrochemical biosensors: A review. Electroanalysis 2005, 17, 7–14.

34. Javey, A.; Guo, J.; Farmer, D.B.; Wang, Q.; Wang, D.W.; Gordon, R.G.; Lundstrom, M.; Dai, H.J. Carbon nanotube field-effect transistors with integrated ohmic contacts and high-k gate dielectrics. Nano Lett.2004, 4, 447–450.

35. Cao, Q.; Han, S.-J.; Tulevski, G.S.; Zhu, Y.; Lu, D.D.; Haensch, W. Arrays of single-walled carbon nanotubes with full surface coverage for high-performance electronics. Nat. Nanotechnol. 2013, 8, 180–186.

36. Franklin, A.D.; Wong, H.S.P.; Lin, A.; Chen, Z.H. Current scaling in aligned carbon nanotube array transistors with local bottom gating. IEEE Electron Device Lett. 2010, 31, 644–646.

37. Cao, J.; Ionescu, A.M. Self-aligned lateral dual-gate suspended-body single-walled carbon nanotube field-effect transistors. Appl. Phys. Lett. 2012, 100, 063103:1–063103:3.

38. Franklin, A.D.; Koswatta, S.O.; Farmer, D.B.; Smith, J.T.; Gignac, L.; Breslin, C.M.; Han, S.-J.; Tulevski, G.S.; Miyazoe, H.; Haensch, W.; et al. Carbon nanotube complementary wrap-gate transistors. Nano Lett. 2013, 13, 2490–2495.

39. Pimparkar, N.; Alam, M.A. A “bottom-up” redefinition for mobility and the effect of poor tube-tube contact on the performance of CNT nanonet thin-film transistors. IEEE Electron Device Lett. 2008, 29, 1037–1039.

40. Pimparkar, N.; Guo, J.; Alam, M.A. Performance assessment of subpercolating nanobundle network thin-film transistors by an analytical model. IEEE Trans. Electron Devices 2007, 54, 637–644.

41. Raychowdhury, A.; de, V.K.; Kurtin, J.; Borkar, S.Y.; Roy, K.; Keshavarzi, A. Variation tolerance in a multichannel carbon-nanotube transistor for high-speed digital circuits. IEEE Trans. Electron Devices2009, 56, 383–392.

42. Nessim, G.D. Properties, synthesis, and growth mechanisms of carbon nanotubes with special focus on thermal chemical vapor deposition. Nanoscale 2010, 2, 1306–1323.

43. Cao, Q.; Rogers, J.A. Ultrathin films of single-walled carbon nanotubes for electronics and sensors: A review of fundamental and applied aspects. Adv. Mater. 2009, 21, 29–53.

44. Islam, A.E.; Du, F.; Ho, X.; Jin, S.H.; Dunham, S.; Rogers, J.A. Effect of variations in diameter and density on the statistics of aligned array carbon-nanotube field effect transistors. J. Appl. Phys. 2012,111, 054511:1–054511:7.

45. Salamat, S.; Ho, X.N.; Rogers, J.A.; Alam, M.A. Intrinsic performance variability in aligned array CNFETs. IEEE Trans. Nanotechnol. 2011, 10, 439–444.

46. Sarker, B.K.; Shekhar, S.; Khondaker, S.I. Semiconducting enriched carbon nanotube aligned arrays of tunable density and their electrical transport properties. ACS Nano 2011, 5, 6297–6305.

47. Kocabas, C.; Kim, H.-S.; Banks, T.; Rogers, J.A.; Pesetski, A.A.; Baumgardner, J.E.; Krishnaswamy, S.V.; Zhang, H. Radio frequency analog electronics based on carbon nanotube transistors. Proc. Natl. Acad. Sci. USA 2008, 105, 1405–1409.

48. Hong, S.W.; Banks, T.; Rogers, J.A. Improved density in aligned arrays of single-walled carbon nanotubes by sequential chemical vapor deposition on quartz. Adv. Mater. 2010, 22, 1826–1830.

49. Xiao, J.L.; Dunham, S.; Liu, P.; Zhang, Y.W.; Kocabas, C.; Moh, L.; Huang, Y.G.; Hwang, K.C.; Lu, C.; Huang, W.; et al. Alignment controlled growth of single-walled carbon nanotubes on quartz substrates. Nano Lett. 2009, 9, 4311–4319.

50. Arnold, M.S.; Green, A.A.; Hulvat, J.F.; Stupp, S.I.; Hersam, M.C. Sorting carbon nanotubes by electronic structure using density differentiation. Nat. Nanotechnol. 2006, 1, 60–65.

51. Heller, D.A.; Mayrhofer, R.M.; Baik, S.; Grinkova, Y.V.; Usrey, M.L.; Strano, M.S. Concomitant length and diameter separation of single-walled carbon nanotubes. J. Am. Chem. Soc. 2004, 126, 14567–14573.

52. Zheng, M.; Jagota, A.; Strano, M.S.; Santos, A.P.; Barone, P.; Chou, S.G.; Diner, B.A.; Dresselhaus, M.S.; McLean, R.S.; Onoa, G.B.; et al. Structure-based carbon nanotube sorting by sequence-dependent DNA assembly. Science 2003, 302, 1545–1548.

53. Stokes, P.; Khondaker, S.I. High quality solution processed carbon nanotube transistors assembled by dielectrophoresis. Appl. Phys. Lett. 2010, 96, 083110:1–083110:3.

54. Jin, S.H.; Dunham, S.N.; Song, J.; Xie, X.; Kim, J.-H.; Lu, C.; Islam, A.; Du, F.; Kim, J.; Felts, J.; et al. Using nanoscale thermocapillary flows to create arrays of purely semiconducting single-walled carbon nanotubes. Nat. Nanotechnol. 2013, 8, 347–355.

55. Shulaker, M.M.; Wei, H.; Patil, N.; Provine, J.; Chen, H.Y.; Wong, H.S.P.; Mitra, S. Linear increases in carbon nanotube density through multiple transfer technique. Nano Lett. 2011, 11, 1881–1886.

56. Wang, C.; Ryu, K.; de Arco, L.G.; Badmaev, A.; Zhang, J.; Lin, X.; Che, Y.; Zhou, C. Synthesis and device applications of high-density aligned carbon nanotubes using low-pressure chemical vapor deposition and stacked multiple transfer. Nano Res. 2010, 3, 831–842.

57. Kim, W.; Javey, A.; Tu, R.; Cao, J.; Wang, Q.; Dai, H.J. Electrical contacts to carbon nanotubes down to 1 nm in diameter. Appl. Phys. Lett. 2005, 87, 173101:1–173101:3.

58. Tseng, Y.C.; Phoa, K.; Carlton, D.; Bokor, J. Effect of diameter variation in a large set of carbon nanotube transistors. Nano Lett. 2006, 6, 1364–1368.

59. Chen, Z.H.; Appenzeller, J.; Knoch, J.; Lin, Y.M.; Avouris, P. The role of metal-nanotube contact in the performance of carbon nanotube field-effect transistors. Nano Lett. 2005, 5, 1497–1502.

60. Nosho, Y.; Ohno, Y.; Kishimoto, S.; Mizutani, T. n-Type carbon nanotube field-effect transistors fabricated by using Ca contact electrodes. Appl. Phys. Lett. 2005, 86, 073105:1–073105:3.

61. Wang, S.; Zeng, Q.S.; Yang, L.J.; Zhang, Z.Y.; Wang, Z.X.; Pei, T.A.; Ding, L.; Liang, X.L.; Gao, M.; Li, Y.; et al. High-performance carbon nanotube light-emitting diodes with Asymmetric contacts. Nano Lett. 2011, 11, 23–29.

62. Wei, H.; Chen, H.-Y.; Liyanage, L.; Wong, H.S.P.; Mitra, S. Air-Stable Technique for Fabricating n-Type Carbon Nanotube FETs. In Proceedings of the IEEE International Electron Devices Meeting, Washington DC, USA, 5–7 December, 2011; pp. 505–508.

63. Franklin, A.D.; Chen, Z.H. Length scaling of carbon nanotube transistors. Nat. Nanotechnol. 2010, 5, 858–862.

64. Franklin, A.D.; Luisier, M.; Han, S.-J.; Tulevski, G.; Breslin, C.M.; Gignac, L.; Lundstrom, M.S.; Haensch, W. Sub-10 nm carbon nanotube transistor. Nano Lett. 2012, 12, 758–762.

65. Lin, Y.M.; Appenzeller, J.; Avouris, P. Novel Carbon Nanotube FET Design with Tunable Polarity. In Proceedings of the IEEE International Electron Devices Meeting, SanFrancisco, CA, 13–15 December, 2004; pp. 687–690.

66. Sze, S.M. Physics of Semiconductor Devices, 2nd ed.; John Wiley & Sons: Hoboken, NY, USA, 1981.

67. Ho, X.; Ye, L.; Rotkin, S.V.; Cao, Q.; Unarunotai, S.; Salamat, S.; Alam, M.A.; Rogers, J.A. Scaling properties in transistors that use aligned arrays of single-walled carbon nanotubes. Nano Lett. 2010, 10, 499–503.

68. Deshpande, V.V.; Chandra, B.; Caldwell, R.; Novikov, D.S.; Hone, J.; Bockrath, M. Mott insulating state in ultraclean carbon nanotubes. Science 2009, 323, 106–110.

69. Charlier, J.C.; Blase, X.; Roche, S. Electronic and transport properties of nanotubes. Rev. Mod. Phys.2007, 79, 677–732.

70. Wahab, M.A.; Jin, S.H.; Islam, A.E.; Kim, J.; Kim, J.-H.; Yeo, W.-H.; Lee, D.J.; Chung, H.U.; Rogers, J.A.; Alam, M.A. Electrostatic dimension of aligned-array carbon nanotube field-effect transistors. ACS Nano 2013, 7, 1299–1308.

71. Rutherglen, C.; Jain, D.; Burke, P. Nanotube electronics for radiofrequency applications. Nat. Nanotechnol. 2009, 4, 811–819.

72. Ryu, K.; Badmaev, A.; Wang, C.; Lin, A.; Patil, N.; Gomez, L.; Kumar, A.; Mitra, S.; Wong, H.S.P.; Zhou, C.W. CMOS-analogous wafer-scale nanotube-on-insulator approach for submicrometer devices and integrated circuits using aligned nanotubes. Nano Lett. 2009, 9, 189–197.

73. Ishikawa, F.N.; Chang, H.K.; Ryu, K.; Chen, P.C.; Badmaev, A.; de Arco, L.G.; Shen, G.Z.; Zhou, C.W. Transparent electronics based on transfer printed aligned carbon nanotubes on rigid and flexible substrates. ACS Nano 2009, 3, 73–79.

74. Feng, C.; Liu, K.; Wu, J.S.; Liu, L.; Cheng, J.S.; Zhang, Y.Y.; Sun, Y.H.; Li, Q.Q.; Fan, S.S.; Jiang, K.L. Flexible, stretchable, transparent conducting films made from superaligned carbon nanotubes. Adv. Funct. Mater. 2010, 20, 885–891.

75. Ago, H.; Imamoto, K.; Ishigami, N.; Ohdo, R.; Ikeda, K.-I.; Tsuji, M. Competition and cooperation between lattice-oriented growth and step-templated growth of aligned carbon nanotubes on sapphire.Appl. Phys. Lett. 2007, 90, 123112:1–123112:3.

76. Ismach, A.; Segev, L.; Wachtel, E.; Joselevich, E. Atomic-step-templated formation of single wall carbon nanotube patterns. Angew. Chem. Int. Ed. 2004, 43, 6140–6143.

77. Kocabas, C.; Kang, S.J.; Ozel, T.; Shim, M.; Rogers, J.A. Improved synthesis of aligned arrays of single-walled carbon nanotubes and their implementation in thin film type transistors. J. Phys. Chem. C 2007,111, 17879–17886.

78. Huang, H.J.; Maruyama, R.; Noda, K.; Kajiura, H.; Kadono, K. Preferential destruction of metallic single-walled carbon nanotubes by laser irradiation. J. Phys. Chem. B 2006, 110, 7316–7320.

79. Collins, P.C.; Arnold, M.S.; Avouris, P. Engineering carbon nanotubes and nanotube circuits using electrical breakdown. Science 2001, 292, 706–709.

80. Chen, C.X.; Yan, L.J.; Kong, E.S.W.; Zhang, Y.F. Ultrasonic nanowelding of carbon nanotubes to metal electrodes. Nanotechnology 2006, 17, 2192–2197.

81. Perebeinos, V.; Avouris, P. Impact excitation by hot carriers in carbon nanotubes. Phy. Rev. B 2006, 74.

82. Liao, A.; Zhao, Y.; Pop, E. Avalanche-induced current enhancement in semiconducting carbon nanotubes. Phys. Rev. Lett. 2008, 101, 256804.

83. Kim, W.; Javey, A.; Vermesh, O.; Wang, O.; Li, Y.M.; Dai, H.J. Hysteresis caused by water molecules in carbon nanotube field-effect transistors. Nano Lett. 2003, 3, 193–198.

84. Polishchuk, I.; Ranade, P.; King, T.J.; Hu, C.M. Dual work function metal gate CMOS technology using metal interdiffusion. IEEE Electron Device Lett. 2001, 22, 444–446.

85. Lim, A.E.J.; Hwang, W.S.; Wang, X.P.; Lai, D.M.Y.; Samudra, G.S.; Kwong, D.L.; Yeo, Y.C. Metal-gate work function modulation using hafnium alloys obtained by the interdiffusion of thin metallic layers.J. Electrochem. Soc. 2007, 154, H309–H313.

86. Franklin, A.D.; Tulevski, G.S.; Han, S.-J.; Shahrjerdi, D.; Cao, Q.; Chen, H.-Y.; Wong, H.S.P.; Haensch, W. Variability in carbon nanotube transistors: Improving device-to-device consistency. ACS Nano2012, 6, 1109–1115.

87. Paul, B.C.; Fujita, S.; Okajima, M.; Lee, T.H.; Wong, H.S.P.; Nishi, Y. Impact of a process variation on nanowire and nanotube device performance. IEEE Trans. Electron Devices 2007, 54, 2369–2376.

88. Wong, H.S.P.; Mitra, S.; Akinwande, D.; Beasley, C.; Chai, Y.; Chen, H.-Y.; Chen, X.; Close, G.; Deng, J.; Hazeghi, A.; et al. Carbon Nanotube Electronics—Materials, Devices, Circuits, Design, Modeling, and Performance Projection. In Proceedings of the IEEE International Electron Devices Meeting, Washington DC, USA, 5–7 December, 2011; pp. 501–504.

89. Balijepalli, A.; Sinha, S.; Cao, Y. Compact Modeling of Carbon Nanotube Transistor for Early Stage Process-Design Exploration. In Proceedings of the 2007 ACM/IEEE International Symposium on Low Power Electronics and Design (ISLPED), Portland, OR, USA, 27–29 August 2007.

90. Zhang, G.; Qi, P.; Wang, X.; Lu, Y.; Li, X.; Tu, R.; Bangsaruntip, S.; Mann, D.; Zhang, L.; Dai, H. Selective etching of metallic carbon nanotubes by gas-phase reaction. Science 2006, 314, 974–977.

91. Zhang, J.; Patil, N.P.; Hazeghi, A.; Wong, H.S.P.; Mitra, S. Characterization and design of Logic circuits in the presence of carbon nanotube density variations. IEEE Trans. Comput. Aided Des. Integr. Circuits Syst. 2011, 30, 1103–1113.

92. Boslaugh, S.; Watters, P.A. Safari Tech Books Online. In Statistics in a Nutshell; O'Reilly Media: Sebastopol, CA, USA, 2008.

93. Sun, Y.; Tuleski, G.; Han, S.-J.; Haensch, W.; Chen, Z. Improve Variability in Carbon Nanotube FETs by Scaling. In Proceedings

of the Device Research Conference, NotreDame, IN, USA, 21–23 June, 2010; pp. 283–284.

94. Jin, S.H.; Islam, A.E.; Kim, T.-I.; Kim, J.-H.; Alam, M.A.; Rogers, J.A. Sources of hysteresis in carbon nanotube field-effect transistors and their elimination via methylsiloxane encapsulants and optimized growth. Adv. Funct. Mater. 2012, 22, 2276–2284.

95. Qu, M.; Qiu, Z.-J.; Zhang, Z.-B.; Li, H.; Li, J.; Zhang, S.-L. Charge-injection-induced time decay in carbon nanotube network-based FETs. IEEE Electron Device Lett. 2010, 31, 1098–1100.

96. Peng, N.; Zhang, Q.; Yuan, S.; Li, H.; Tian, J.; Chan, L. Current instability of carbon nanotube field effect transistors. Nanotechnology 2007, 18.

97. Ellison, M.D.; Crotty, M.J.; Koh, D.; Spray, R.L.; Tate, K.E. Adsorption of NH3 and NO2 on single-walled carbon nanotubes. J. Phys. Chem. B 2004, 108, 7938–7943.

98. Yao, Z.; Postma, H.W.C.; Balents, L.; Dekker, C. Carbon nanotube intramolecular junctions. Nature1999, 402, 273–276.

99. Robinson, J.A.; Snow, E.S.; Badescu, S.C.; Reinecke, T.L.; Perkins, F.K. Role of defects in single-walled carbon nanotube chemical sensors. Nano Lett. 2006, 6, 1747–1751.

100. Zhou, C.W.; Kong, J.; Yenilmez, E.; Dai, H.J. Modulated chemical doping of individual carbon nanotubes. Science 2000, 290, 1552–1555.

101. Zhuravlev, L.T. The surface chemistry of amorphous silica. Zhuravlev model. Colloids Surfaces Physicochem. Eng. Asp. 2000, 173, 1–38.

102. Chakrapani, V.; Anderson, A.B.; Wolter, S.D.; Stoner, B.R.; Sumanasekera, G.U.; Angus, J.C. Charge transfer equilibria between diamond and an aqueous oxygen electrochemical redox couple. Science2007, 318, 1424–1430.

103. McGill, S.A.; Rao, S.G.; Manandhar, P.; Xiong, P.; Hong, S. High-performance, hysteresis-free carbon nanotube field-effect transistors via directed assembly. Appl. Phys. Lett. 2006, 89, 163123:1–163123:3.

104. Aguirre, C.M.; Levesque, P.L.; Paillet, M.; Lapointe, F.; St-Antoine, B.C.; Desjardins, P.; Martel, R. The role of the oxygen/water

redox couple in suppressing electron conduction in field-effect transistors.Adv. Mater. 2009, 21.

105. Levesque, P.L.; Sabri, S.S.; Aguirre, C.M.; Guillemette, J.; Siaj, M.; Desjardins, P.; Szkopek, T.; Martel, R. Probing charge transfer at surfaces using graphene transistors. Nano Lett. 2011, 11, 132–137.

106. Weitz, R.T.; Zschieschang, U.; Effenberger, F.; Klauk, H.; Burghard, M.; Kern, K. High-performance carbon nanotube field effect transistors with a thin gate dielectric based on a self-assembled monolayer. Nano Lett. 2007, 7, 22–27.

107. Hur, S.H.; Yoon, M.H.; Gaur, A.; Shim, M.; Facchetti, A.; Marks, T.J.; Rogers, J.A. Organic nanodielectrics for low voltage carbon nanotube thin film transistors and complementary logic gates.J. Am. Chem. Soc. 2005, 127, 13808–13809.

108. Reisman, A.; Williams, C.K.; Maldonado, J.R. Generation and annealing of defects in silicon dioxide. J. Appl. Phys. 1987, 62, 868–874.

109. Alam, M.A. SILC as a measure of trap generation and predictor of T-BD in ultrathin oxides. IEEE Trans. Electron Devices 2002, 49, 226–231.

110. Solomon, P. Breakdown in silicon-oxide. J. Vac. Sci. Technol. 1977, 14, 1122–1130.

111. Ribes, G.; Mitard, J.; Denais, M.; Bruyere, S.; Monsieur, F.; Parthasarathy, C.R.; Vincent, E.; Ghibaudo, G. Review on high-k dielectrics reliability issues. IEEE Trans. Device Mater Reliab. 2005, 5, 5–19.

112. Lenahan, P.M.; Dressendorfer, P.V. Hole traps and trivalent silicon centers in metal-oxide silicon devices. J. Appl. Phys. 1984, 55, 3495–3499.

113. Lenahan, P.M. Dominating defects in the MOS system: Pb and E' centers. In Defects in Microelectronic Materials and Devices; Fleetwood, S.P.D., Schrimpf, R.D., Eds.; CRC Press: Taylor and Francis Group, Boca Raton, FL, USA, 2009; pp. 163–214.

114. Skuja, L. Optically active oxygen-deficiency-related centers in amorphous silicon dioxide. J. Non Cryst. Solids 1998, 239, 16–48.

115. Spinelli, A.S.; Lacaita, A.L.; Rigamonti, M.; Ielmini, D.; Ghidini, G. Separation of electron and hole traps by transient current analysis. Microelectron. Eng. 1999, 48, 151–154.

116. Hongo, H.; Nihey, F.; Yorozu, S. Relationship between carbon nanotube density and hysteresis characteristics of carbon nanotube random network-channel field effect transistors. J. Appl. Phys. 2010,107, 094501:1–094501:9.

117. Di Bartolomeo, A.; Rinzan, M.; Boyd, A.K.; Yang, Y.; Guadagno, L.; Giubileo, F.; Barbara, P. Electrical properties and memory effects of field-effect transistors from networks of single- and double-walled carbon nanotubes. Nanotechnology 2010, 21.

118. Memming, R. Semiconductor Electrochemistry; Wiley-VCH: Weinheim, NY, USA, 2001.

119. Shockley, W.; Read, W.T. Statistics of the recombinations of holes and electrons. Phys. Rev. 1952, 87, 835–842.

120. Chang, M.H.; Zhang, J.F.; Zhang, W.D. Assessment of capture cross sections and effective density of electron traps generated in silicon dioxides. IEEE Trans. Electron Devices 2006, 53, 1347–1354.

121. Hu, P.; Zhang, C.; Fasoli, A.; Scardaci, V.; Pisana, S.; Hasan, T.; Robertson, J.; Milne, W.I.; Ferrari, A.C. Hysteresis suppression in self-assembled single-wall nanotube field effect transistors. Phys. ELowDimens. Syst. Nanostruct. 2008, 40, 2278–2282.

122. Kim, S.K.; Xuan, Y.; Ye, P.D.; Mohammadi, S.; Back, J.H.; Shim, M. Atomic layer deposited Al2O3 for gate dielectric and passivation layer of single-walled carbon nanotube transistors. Appl. Phys. Lett.2007, 90, 163108:1–163108:3.

123. Kirton, M.J.; Uren, M.J. Noise in solid-state microstructures—A new perspective on individual defects, interface states and low-frequency (1/f) noise. Adv. Phys. 1989, 38, 367–468.

124. Vandamme, L.K.J. Noise as a diagnostic tool for quality and reliability of electronic devices. IEEE Trans. Electron Devices 1994, 41, 2176–2187.

125. Vandamme, L.K.J.; Li, X.S.; Rigaud, D. 1/f Noise in MOS Devices, Mobility or Number Fluctuations.IEEE Trans. Electron Devices 1994, 41, 1936–1945.

126. Weissman, M.B. 1/f noise and other slow, nonexponential kinetics in condensed matter. Rev. Mod. Phys. 1988, 60, 537–571.

127. Weissman, M.B. Low-frequency noise as a tool to study disordered materials. Annu. Rev. Mater. Sci.1996, 26, 395–429.

128. Kurata, H.; Otsuga, K.; Kotabe, A.; Kajiyama, S.; Osabe, T.; Sasago, Y.; Narumi, S.; Tokami, K.; Kamohara, S.; Tsuchiya, O. Random telegraph signal in flash memory: Its impact on scaling of multilevel flash memory beyond the 90-nm node. IEEE J. SolidState Circuits 2007, 42, 1362–1369.

129. Ghetti, A.; Compagnoni, C.M.; Biancardi, F.; Lacaita, A.L.; Beltrami, S.; Chiavarone, L.; Spinelli, A.S.; Visconti, A. Scaling trends for random telegraph noise in deca-nanometer Flash memories. In Proceedings of the IEEE International Electron Devices Meeting Technical Digest, SanFrancisco, CA, 15–17 December, 2008; pp. 835–838.

130. Gierkink, S.L.J.; Klumperink, E.A.M.; van der Wel, A.P.; Hoogzaad, G.; van Tuijl, E.; Nauta, B. Intrinsic 1/f device noise reduction and its effect on phase noise in CMOS ring oscillators. IEEE J. SolidState Circuits 1999, 34, 1022–1025.

131. Cao, Q.; Xia, M.G.; Kocabas, C.; Shim, M.; Rotkin, S.V.; Rogers, J.A. Gate capacitance coupling of singled-walled carbon nanotube thin-film transistors. Appl. Phys. Lett. 2007, 90, 023516:1–023516:3.

132. Rosenblatt, S.; Yaish, Y.; Park, J.; Gore, J.; Sazonova, V.; McEuen, P.L. High performance electrolyte gated carbon nanotube transistors. Nano Lett. 2002, 2, 869–872.

133. Vitusevich, S.A.; Sydoruk, V.A.; Petrychuk, M.V.; Danilchenko, B.A.; Klein, N.; Offenhaeusser, A.; Ural, A.; Bosman, G. Transport properties of single-walled carbon nanotube transistors after gamma radiation treatment. J. Appl. Phys. 2010, 107, 063701:1–063701:6.

134. Lin, Y.-M.; Appenzeller, J.; Knoch, J.; Chen, Z.; Avouris, P. Low-frequency current fluctuations in individual semiconducting single-wall carbon nanotubes. Nano Lett. 2006, 6, 930–936.

135. Collins, P.G.; Fuhrer, M.S.; Zettl, A. 1/f noise in carbon nanotubes. Appl. Phys. Lett. 2000, 76, 894–896.

136. Snow, E.S.; Novak, J.P.; Lay, M.D.; Perkins, F.K. 1/f noise in single-walled carbon nanotube devices.Appl. Phys. Lett. 2004, 85, 4172–4174.

137. Liu, F.; Wang, K.L.; Li, C.; Zhou, C. Study of random telegraph signals in single-walled carbon nanotube field effect transistors. IEEE Trans. Nanotechnol. 2006, 5, 441–445.

138. Vitusevich, S.; Gasparyan, F. Low-Frequency Noise Spectroscopy at Nanoscale: Carbon Nanotube Materials and Devices. Available online: http://cdn.intechweb.org/pdfs/17294.pdf (accessed on 1 July 2013).

139. Veksler, D.; Bersuker, G.; Rumyantsev, S.; Shur, M.; Park, H.; Young, C.; Lim, K.Y.; Taylor, W.; Jammy, R.; IEEE. Understanding noise measurements in MOSFETs: The role of traps structural relaxation. In Proceedings of the International Reliability Physics Symposium, Anaheim, CA, 2–6 May, 2010; pp. 73–79.

140. Hooge, F.N. 1/F noise sources. IEEE Trans. Electron Devices 1994, 41, 1926–1935.

141. Liu, F.; Bao, M.; Wang, K.L.; Zhang, D.; Zhou, C. Coulomb attractive random telegraph signal in a single-walled carbon nanotube. Phys. Rev. B 2006, 74.

142. Liu, F.; Wang, K.L. Correlated random telegraph signal and low-frequency noise in carbon nanotube transistors. Nano Lett. 2008, 8, 147–151.

143. Liu, F.; Wang, K.L.; Zhang, D.; Zhou, C. Random telegraph signals and noise behaviors in carbon nanotube transistors. Appl. Phys. Lett. 2006, 89, 243101:1–243101:3.

144. Chan, J.; Kidd, D.; Burke, B.; Harriott, L.; Williams, K. Noise analysis of carbon nanotube field effect transistors irradiated by electron beam. J. Vac. Sci. Technol. B 2010, 28, C6P66–C6P69.

145. Cress, C.D.; McMorrow, J.J.; Robinson, J.T.; Friedman, A.L.; Hughes, H.L.; Weaver, B.D.; Landi, B.J. Total ionizing dose-hardened carbon nanotube thin-film transistors with silicon oxynitride gate dielectrics. MRS Commun. 2011, 1, 27–31.

146. Cress, C.D.; McMorrow, J.J.; Robinson, J.T.; Friedman, A.L.; Landi, B.J. Radiation effects in single-walled carbon nanotube thin-film-transistors. IEEE Trans. Nucl. Sci. 2010, 57, 3040–3045.

147. Hulman, M.; Skakalova, V.; Roth, S.; Kuzmany, H. Raman spectroscopy of single-wall carbon nanotubes and graphite irradiated by gamma rays. J. Appl. Phys. 2005, 98, 024311:1–024311:5.

148. Rossi, J.E.; Cress, C.D.; Helenic, A.R.; Schauerman, C.M.; DiLeo, R.A.; Cox, N.D.; Messenger, S.R.; Weaver, B.D.; Hubbard, S.M.; Landi, B.J. Ion irradiation of electronic-type-separated single wall carbon nanotubes: A model for radiation effects in nanostructured carbon. J. Appl. Phys. 2012, 112, 034314:1–034314:11.

149. Tang, X.W.; Yang, Y.; Kim, W.; Wang, Q.; Qi, P.F.; Dai, H.J.; Xing, L. Measurement of ionizing radiation using carbon nanotube field effect transistor. Phys. Med. Biol. 2005, 50, N23–N31.

150. Cress, C.D.; Schauerman, C.M.; Landi, B.J.; Messenger, S.R.; Raffaelle, R.P.; Walters, R.J. Radiation effects in single-walled carbon nanotube papers. J. Appl. Phys. 2010, 107, 014316:1–014316:5.

151. Skakalova, V.; Woo, Y.S.; Osvath, Z.; Biro, L.P.; Roth, S. Electron transport in Ar+-irradiated single wall carbon nanotubes. Phys. Status Solidi B 2006, 243, 3346–3350.

152. Dresselhaus, M.S.; Dresselhaus, G.; Saito, R.; Jorio, A. Raman spectroscopy of carbon nanotubes. Phys. Rep. Rev. Sect. Phys. Lett. 2005, 409, 47–99.

153. Buchowicz, G.; Stone, P.R.; Robinson, J.T.; Cress, C.D.; Beeman, J.W.; Dubon, O.D. Correlation between structure and electrical transport in ion-irradiated graphene grown on Cu foils. Appl. Phys. Lett. 2010, 98, 032102:1–032102:3.

154. Hong, W.-K.; Lee, C.; Nepal, D.; Geckeler, K.E.; Shin, K.; Lee, T. Radiation hardness of the electrical properties of carbon nanotube network field effect transistors under high-energy proton irradiation.Nanotechnology 2006, 17, 5675–5680.

155. Hughes, H.L.; Benedetto, J.M. Radiation effects and hardening of MOS technology: Devices and circuits. IEEE Trans. Nucl. Sci. 2003, 50, 500–521.

156. Islam, A.E.; Kufluoglu, H.; Varghese, D.; Mahapatra, S.; Alam, M.A. Recent issues in negative bias temperature instability: Initial degradation, field-dependence of interface trap generation, hole

trapping effects, and relaxation. IEEE Trans. Electron Devices 2007, 54, 2143–2154.

157. Schroder, D.K.; Babcock, J.A. Negative bias temperature instability: Road to cross in deep submicron silicon semiconductor manufacturing. J. Appl. Phys. 2003, 94, 1–18.

158. Gu, G.; Kane, M.G.; Mau, S.C. Reversible memory effects and acceptor states in pentacene-based organic thin-film transistors. J. Appl. Phys. 2007, 101, 014504:1–014504:9.

159. Liu, J.; Wang, C.; Tu, X.; Liu, B.; Chen, L.; Zheng, M.; Zhou, C. Chirality-controlled synthesis of single-wall carbon nanotubes using vapour-phase epitaxy. Nat. Commun. 2012, 3.

160. Wilk, G.D.; Wallace, R.M.; Anthony, J.M. High-kappa gate dielectrics: Current status and materials properties considerations. J. Appl. Phys. 2001, 89, 5243–5275.

9

LOCALLY-ACTUATED GRAPHENE-BASED NANO-ELECTRO-MECHANICAL SWITCH

Jian Sun [1,2,*], Manoharan Muruganathan [1], Nozomu Kanetake [1] and Hiroshi Mizuta [1,3]

[1]School of Material Science, Japan Advanced Institute of Science and Technology, Nomi 923-1211, Japan

[2]Advanced Device Laboratory, RIKEN, Wako 351-0198, Japan

[3]Nanoelectronics and Nanotechnologies Research Group, Faculty of Physical Sciences and Engineering, University of Southampton, High Field, Southampton SO17 1BJ, UK

ABSTRACT

The graphene nano-electro-mechanical switches are promising components due to their outstanding switching performance. However, most of the reported devices suffered from a large actuation voltages, hindering them from the integration in the conventional complementary metal-oxide-semiconductor (CMOS) circuit. In this work, we demonstrated the graphene nano-electro-mechanical switches with the local actuation electrode via conventional nanofabrication techniques. Both cantilever-type and double-clamped beam switches were fabricated. These devices exhibited the sharp switching, reversible operation cycles, high on/off ratio, and a low actuation voltage of below 5 V, which were compatible with the CMOS circuit requirements.

Keywords: graphene; nano-electro-mechanical system; nano-electro-mechanical switch; nanofabrication

1. INTRODUCTION

Nano-electro-mechanical (NEM) switches, utilizing electrostatic forces to mechanically deflect the active element into physical contact with an electrode, are of great interest for future logic devices, relays and sensors [1]. Graphene has an ultra-high Young's modulus of 1 TPa, making it a promising candidate for future NEM applications. The graphene NEM contact switches showed minimized electrical leakage, sharp switching response, low actuation voltage, and a high on/off ratio [2,3,4,5,6,7]. In most of the reported works, the heavily-doped silicon substrate has been utilized as the actuation electrode. However, as a consequence, the graphene NEM switch suffers from the relatively large pull-in voltage, normally greater than 10 V [8]. It is mainly ascribed to the thick, inevitable dielectric gap in additional to a pure air gap in the practical globally-actuated graphene switch. Hence, there is a demand to include the local actuation electrode in order to reduce the pull-in voltage. However, the reported fabrication procedure is complicated, i.e., burying a local bottom gate under the graphene, and not compatible with the standard thin-film bottom-up fabrication process techniques [9]. Moreover, it is known that the cantilever-type suspended structure has a smaller spring constant, i.e., mechanical strength, than the double-clamped beam. A graphene cantilever, in principle, could much reduce the pull-in voltage based on the proper geometric design. However, we noticed just a few experimental studies on graphene cantilevers [10], and only one of them reported the cantilever-type graphene switch device [9], as its fabrication is much more challenging compared with the double-clamped beam.

In this work, we demonstrated the simple procedure to fabricate graphene NEM switches with both double-clamped beam and cantilever-type moving elements. By this method, the local top electrodes can be introduced for actuation. The fabricated devices were characterized with a two-terminal configuration, showing low pull-in voltages of less than 5 V. The results demonstrated the possibility of integrating such NEM switches with a conventional complementary metal-oxide-semiconductor (CMOS) circuit for future applications.

2. EXPERIMENTAL DETAILS

2.1. Switch Fabrication

The fabrication process started with the mechanically-exfoliated monolayer graphene flakes on the p-doped silicon substrate covered with 300 nm thermal SiO_2. The possible graphene flakes were firstly identified from their contrast to the substrate under an optical microscope. Figure 1a shows an exfoliated graphene flake with the various numbers of layers. Then Raman spectroscopy (HeNe laser: 633 nm excitation) was utilized to verify the actual number of layers of the exfoliated graphene [11]. Monolayer graphene was firstly identified by a larger intensity of the 2D band over the G band (Figure 1b). Moreover, its 2D band can be fitted with a single Lorentzian peak with a full width at a half-maximum of <30 cm^{-1}. For the bilayer graphene, four Lorentzian sub-peaks are fitted to the 2D band. In principle, the exfoliated graphene should be intact from significant lattice defects. The mild D peak observed in the measured spectra are ascribed to the defective edges of the flake near the probed areas.

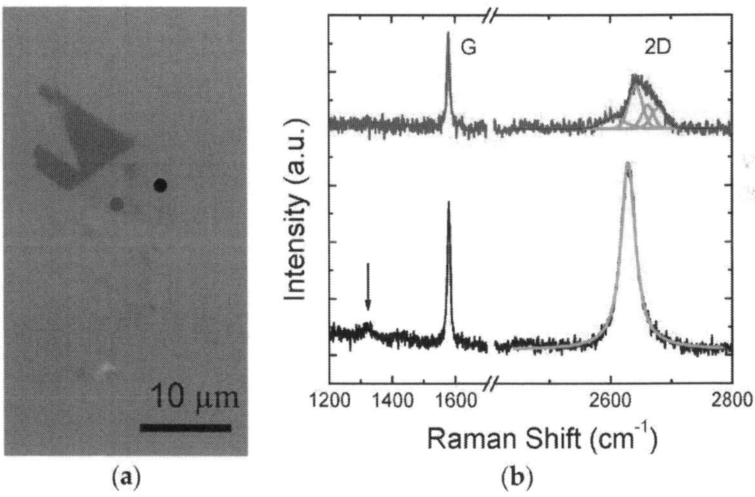

Figure 1. (**a**) Exfoliated graphene flake, black and red dots indicate the monolayer and bilayer regions, respectively; and (**b**) Raman spectra probed at locations marked as two color dots in (**a**). Green lines are the sub-peaks obtained from Lorentzian peak fitting. The arrow indicates the location of D band.

Figure 2 sketches the fabrication process after exfoliation. (a) Firstly, the bottom contacts to graphene flakes were defined with a Cr/Au (5 nm/60 nm) stack following the electron-beam lithography (EBL), evaporation, and lift-off processes; (b) then, hydrogen-silsesquioxane (HSQ) resist was spun and patterned by electron beam lithography (EBL) into the desired shapes acting as the etching masks for the next step; (c) graphene was patterned by transferring the shapes from capping HSQ masks in the oxygen plasma environment in a reactive ion ether; (d) the SiO_2 layer was evaporated and capped onto graphene/HSQ as a sacrificial layer; (e) the top actuation electrode was defined with Cr/Au (5 nm/200 nm); and (f) etch all of the oxide layers (HSQ, sacrificial SiO_2, and SiO_2 substrate) in the buffered hydrofluoric acid (1:5) to release graphene from the supporting substrate. Finally, the device was dried in a critical point drier to prevent surface tension-induced collapse of the suspended graphene. Figure 3 shows a typical fabricated switch.

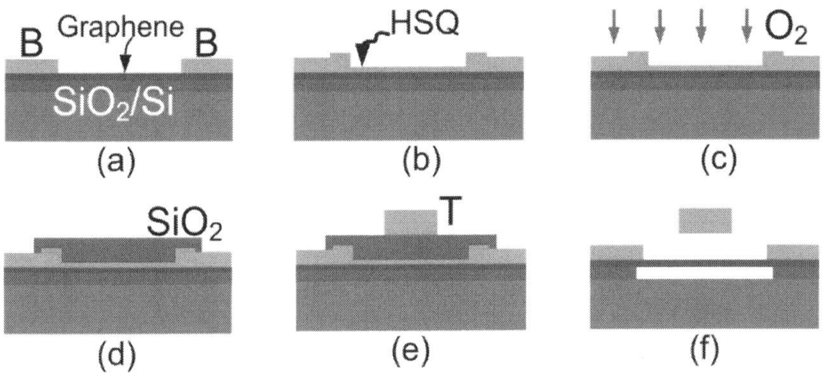

Figure 2. Schematics of fabrication procedure of the graphene NEM switch with local top actuation electrode. T and B denote top and bottom electrodes, respectively. (**a**) Graphene exfoliation and bottom electrodes definition; (**b**) HSQ spin coating and pattering; (**c**) graphene etching in oxygen plasma; (**d**) definition of SiO_2 sacrificial layer; (**e**) definition of top electrode; (**f**) releasing graphene in HF.

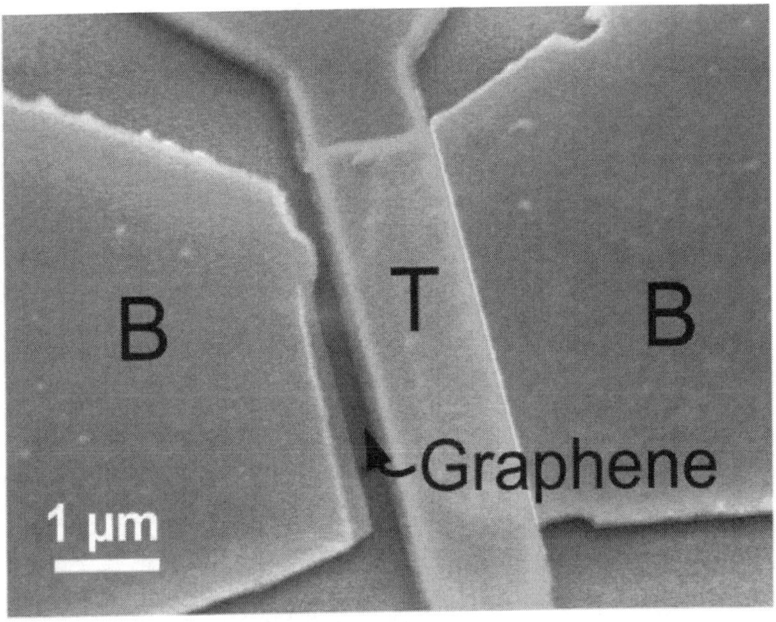

Figure 3. False color SEM image of a double-clamped beam graphene switch as-fabricated. T and B denote top and bottom electrodes, respectively.

2.2. Device Characterization

All of the fabricated graphene switches were characterized in the vacuum condition (~0.1 Pa) to reduce the impacts from ambient, such as the molecular absorption and capillary force due to moisture. The two-terminal configuration is used to investigate switch performance (lower inset of Figure 4), where the top electrode works as both the current drain and actuation electrode. A variable voltage V_{tb} was applied between top and bottom electrodes and swept from 0 V to higher voltages, which could electrostatically deflect the graphene upwards to the top electrode, while the current between two electrodes I_{tb} was monitored.

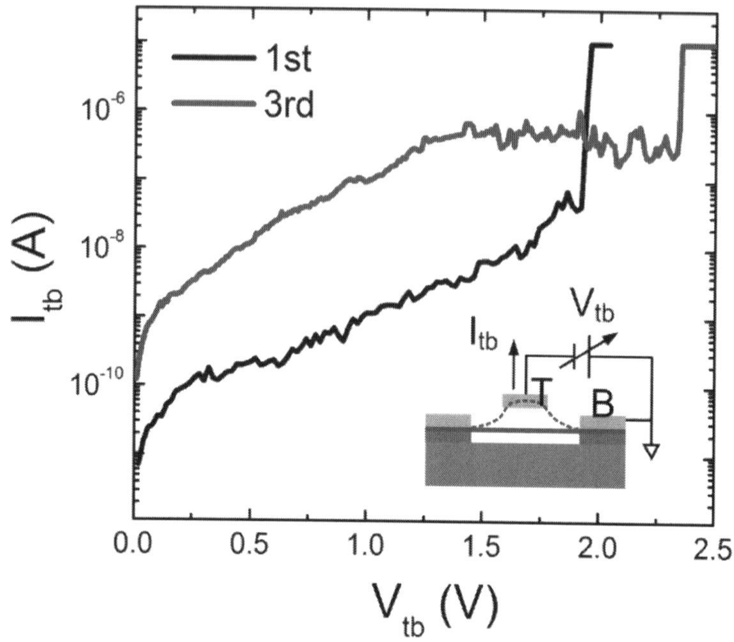

Figure 4. Electrical characterization of the switching performance of a double-clamped beam switch. The inset shows the two-terminal measurement configuration.

3. RESULTS AND DISCUSSIONS

3.1. Double-Clamped Beam Switch

Figure 3 shows the scanning electron microscope (SEM) image of a fabricated switch with a double-clamped monolayer graphene beam and top actuation electrode. The graphene beam has a length l of 1 μm and width w of 2 μm. An air gap between the graphene and top electrode is clearly noticed. The gap thickness g_0 is almost defined by the total thickness of HSQ and SiO_2 sacrificial layer, which is ~180 nm in this device.

At the initial stage, an open circuit was expected between the top and bottom electrode, namely, the switch-off status. Hence, at the beginning of the V_{tb} sweeping, only leakage current of a pico-ampere level was measured. At ~1.92 V, we noticed an abrupt increase of current to the compliance value. This strongly indicates that graphene was physically pulled into the top electrode, as the electrostatic force applied by the top electrode balanced the mechanical restoring force in the graphene beam. In other words, the device was switched on. This actuation voltage is often termed as the pull-in voltage V_{pi}. Comparing to those switches using doped Si substrates as global actuation electrodes [1,8], the much stronger electrostatic force generated by the local top electrode brings about the low V_{pi}, benefiting from the well-controlled pure air gap. Moreover, the steep switching slope of ~15 mV/dec and on/off ratio of $\sim 10^3$ was read from the measurement, highlighting the rapid switching response. Note, a low compliance current of 10 μA was set for this measurement, leading to a much underestimated on current. By assuming the on current in the microampere range, the accrual on/off ratio should be in the order of 10^5.

The pull-in voltage of a double-clamped beam can be theoretically calculated as [8]:

$$V_{pi} = \sqrt{\frac{8kg_0^3}{27\varepsilon Wl}},\ k = 32Ew\left(\frac{t}{l}\right)^3 \qquad (1)$$

where k is the spring constant of the graphene beam, E is the Young's modulus of graphene, t, l, w are the thickness, length, and width of the graphene beam, respectively, ε is permittivity, and W is the overlapping width of top electrode on graphene. With Equation (1) the pull-in voltage of the switch with a clean graphene, which has the thickness t of 0.34 nm, is estimated as 0.70 V. However, the contaminations and adsorbates on graphene are difficult to avoid after fabrication. By assuming a monolayer of water molecularly adsorbed on the graphene surface, it increases the beam thickness to ~0.7 nm; V_{tb} of 2.07 V is then calculated with Equation (1), which is close to the measurement. The current annealing process could be employed to remove the adsorbates [12].

We retracted the actuation voltage application immediately after the observation of pull-in to avoid the joule heating accumulation at the contact interface. This is known as the main reason of device failure, since it triggers the formation of chemical bonds between graphene and the contact metal and causes graphene sticking on contact [6]. Reversible switching operation was realized in this device. However, the slight fluctuation in pull-in voltage was observed; a V_{tb} of 2.3 V was measured in the third operation cycle. It is ascribed to the enhanced leakage current through the degraded substrate (read as the enhanced leakage current background), which effectively reduces the strength of the electric field flux at the graphene surface and, therefore, weakening the electrostatic force.

3.2. Cantilever-Type Switch

Figure 5a shows the SEM image of a few-layer (~10 layers) graphene cantilever switch. An air gap between the graphene and top electrode defined by the sacrificial layer is clearly noticed.

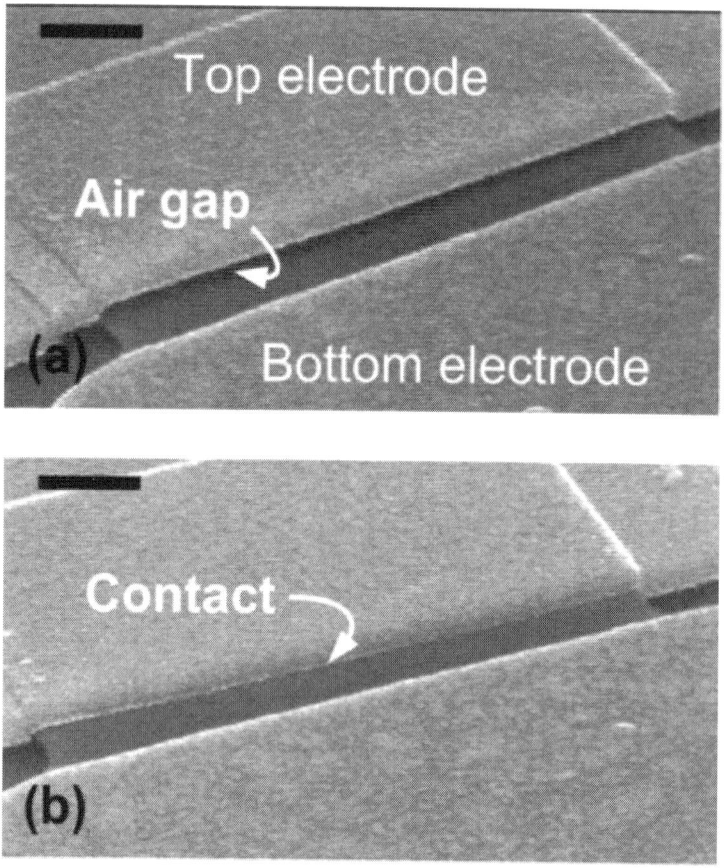

Figure 5. SEM images of a graphene cantilever-type switch with a local top actuation electrode (**a**) before and (**b**) after the switching operation. The initial air gap g_0 between the graphene and the top electrode is about 140 nm. The graphene cantilever is artificially colored in light blue. The scale bars are 500 nm.

A sharp "pull-in" behavior was observed at the V_{pi} of 3.8 V (Figure 6). However, this switch was not reversible after the first operation. The linear I-V response was measured between top and bottom electrode indicating that the graphene contacts to the top electrode (left upper inset of Figure 6). Later, the SEM image confirmed that graphene is stuck on the top electrode without any recognized air gap (Figure 5b).

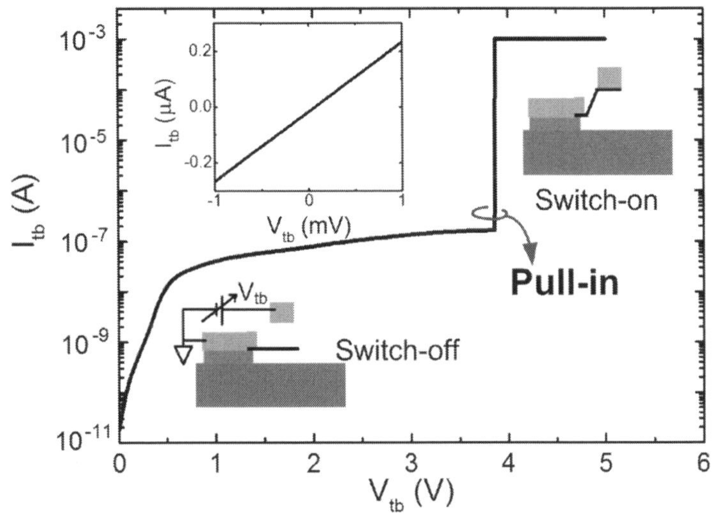

Figure 6. Electrical characterization of the switching performance of a cantilever-type switch. Inset: (**Lower**) two-terminal configuration and switch-off status, (**Right upper**) switch-on status, and (**Left upper**) I-V response after device failure.

As is known, the spring constant k of a cantilever is:

$$k = \frac{2Ew}{3}\left(\frac{t}{l}\right)^3 \tag{2}$$

which is 48 times lower than the double-clamped beam (Equation (1)). Considering the same deflection, the mechanical restoring force is also 48 times weaker than that in a double-clamped beam of the same dimension. However, interfacial interactions between graphene and the actuation electrode, e.g., van der Waals force, chemical bonds, etc., depend only on the material properties. Hence, in the case of a graphene cantilever, the interfacial force is stronger than its mechanical restoring force, and holds the cantilever on the top electrode even after the retraction of actuation voltage. In a future study, in order to have the reversible switching, surface modification is necessary to reduce these interfacial interactions [6].

4. CONCLUSIONS

We demonstrated the graphene NEM switch with local actuation electrode fabricated via simple fabrication methods. Sharp switching is realized at a low voltage of less than 5 V. These outstanding features benefit future applications of graphene NEM devices for high-performance switching components. In the future, the further lowered pull-in voltage can be achieved through the geometric design. More effort will also be devoted to the improvement of device reliability. For instance, contact interface modification will be utilized to prevent graphene from sticking on contact and increase the number of operation cycles.

ACKNOWLEDGMENTS

This work was supported by the Grant-in-Aid for Scientific Research No. 25220904 from Japan Society for the Promotion of Science.

AUTHOR CONTRIBUTIONS

J.S. and H.M. conceived and designed the experiments; J.S. and N.K. performed the experiments; J.S., M.M. and H.M. analyzed the data and discussed the results; J.S., and M.M. wrote the paper.

CONFLICTS OF INTEREST

The authors declare no conflict of interest.

REFERENCES

1. Loh, O.Y.; Espinosa, H.D. Nanoelectromechanical contact switches. Nat. Nanotechnol. 2012, 7, 283–295.

2. Milaninia, K.M.; Baldo, M.A.; Reina, A.; Kong, J. All graphene electromechanical switch fabricated by chemical vapor deposition. Appl. Phys. Lett. 2009, 95, 183105.

3. Kim, S.M.; Song, E.B.; Lee, S.; Seo, S.; Seo, D.H.; Hwang, Y.; Candler, R.; Wang, K.L. Suspended few-layer graphene beam electromechanical switch with abrupt on-off characteristics and minimal leakage current. Appl. Phys. Lett. 2011, 99, 023103.

4. Nagase, M.; Hibino, H.; Kageshima, H.; Yamaguchi, H. Graphene based nanoelectromechanical switch with high On/Off ratio. Appl. Phys. Express 2013, 6, 055101.

5. Li, P.; You, Z.; Haugstad, G.; Cui, T. Graphene fixed-end beam arrays based on mechanical exfoliation.Appl. Phys. Lett. 2011, 98, 253105.

6. Sun, J.; Wang, W.; Muruganathan, M.; Mizuta, H. Low pull-in voltage graphene electromechanical switch fabricated with a polymer sacrificial spacer. Appl. Phys. Lett. 2014, 105, 033103.

7. Sun, J.; Schmidt, M.E.; Muruganathan, M.; Chong, H.M.; Mizuta, H. Large-scale nanoelectromechanical switches based on directly deposited nanocrystalline graphene on insulating substrates. Nanoscale 2016, 8, 6659–6665.

8. Li, P.; You, Z.; Cui, T. Raman spectrum method for characterization of pull-in voltages of graphene capacitive shunt switches. Appl. Phys. Lett. 2012, 101, 263103.

9. Li, P.; You, Z.; Cui, T. Graphene cantilever beams for nano switches. Appl. Phys. Lett. 2012, 101, 093111.

10. Reserbat-Plantey, A.; Marty, L.; Arcizet, O.; Bendiab, N.; Bouchiat, V. A local optical probe for measuring motion and stress in a nanoelectromechanical system. Nat. Nanotechnol. 2012, 7, 151–155.

11. Ferrari, A.C.; Meyer, J.C.; Scardaci, V.; Casiraghi, C.; Lazzeri, M.; Mauri, F.; Piscanec, S.; Jiang, D.; Novoselov, K.S.; Roth, S.; et al. Raman spectrum of graphene and graphene layers. Phys. Rev. Lett.2006, 97, 187401.

12. Moser, J.; Barreiro, A.; Bachtold, A. Current-induced cleaning of graphene. Appl. Phys. Lett. 2007, 91, 163513.

10

IN SITU SYNTHESIS OF REDUCED GRAPHENE OXIDE AND GOLD NANOCOMPOSITES FOR NANOELECTRONICS AND BIOSENSING

Xiaochen Dong[1], Wei Huang[1*], Peng Chen[2]

[1]Key Laboratory for Organic Electronics and Information Displays (KLOEID) and Institute of Advanced Materials (IAM), Nanjing University of Posts and Telecommunications (NUPT), 9 Wenyuan Road, 210046, Nanjing, China.

[2]School of Chemical and Biomedical Engineering, Nanyang Technological University, Singapore, 637457, Singapore.

ABSTRACT

In this study, an in situ chemical synthesis approach has been developed to prepare graphene–Au nanocomposites from chemically reduced graphene oxide (rGO) in aqueous media. UV–Vis absorption, atomic force microscopy, scanning electron microscopy, transmission electron microscopy, and Raman spectroscopy were used to demonstrate the successful attachment of Au nanoparticles to graphene sheets.

Configured as field-effect transistors (FETs), the as-synthesized single-layered rGO-Au nanocomposites exhibit higher hole mobility and conductance when compared to the rGO sheets, promising its applications in nanoelectronics. Furthermore, we demonstrate that the rGO-Au FETs are able to label-freely detect DNA hybridization with high sensitivity, indicating its potentials in nanoelectronic biosensing.

Keywords: Graphene oxide Gold nanoparticles Nanoelectronics Biosensing

INTRODUCTION

Graphene, a single-layer of carbon atoms densely compacted into a two-dimensional honeycomb crystal lattice, has attracted tremendous attention in recent years, because of its unique electronic, optical, thermal, and mechanical properties [1, 2, 3, 4, 5, 6, 7, 8, 9]. It provides a variety of novel applications such as field-effect transistors (FETs) [10, 11], ultrasensitive sensors [12, 13], transparent electrodes [14], and novel nanocomposites [15, 16]. Graphene is particularly advantageous in biosensing due to its large detection area, high charge mobility, low noise, and biocompatibility [17].

The common fabrication methods of graphene sheets include mechanical exfoliation from graphite [18], chemical exfoliation (chemical oxidation of graphite and subsequent reduction of the exfoliated graphite oxide sheets) [19], and epitaxial growth on the surface of silicon carbide crystals or metal substrates [20, 21]. Among these methods, chemical exfoliation is particularly advantageous for low-cost, large-scale, high-yield preparation of graphene sheets. However, the electronic properties of the graphene sheets obtained by this method are not good enough for nanoelectronics. In order to enhance the electrical properties of the chemically reduced GO (rGO) sheets, many approaches have been investigated, such as changing the oxidation methods [22], reduction with different reduction agents or reduction conditions [23, 24], and modification with metal nanoparticles [25].

In this study, we report a simple method for in situ synthesis of rGO-Au nanocomposites. With the assistance of sodium dodecyl sulfate (SDS), the resultant rGO-Au nanocomposites can form stable dispersion in water. Au nanoparticles formed uniformly on the rGO surface. The as-prepared rGO-Au hybrids show significantly higher hole mobility than Au-free rGO sheets. Furthermore, we demonstrate that that field-effect transistors (FETs) based on rGO-Au nanocomposites were able to detect

DNA hybridization with high sensitivity, indicating its promising potentials in printable nanoelectronics and bio-nanoelectronics [26].

EXPERIMENTAL

Materials

Nature graphite flakes were obtained from NGS, Germany. Also, 98% H_2SO_4, 30% H_2O_2, potassium permanganate ($KMnO_4$), and other reagents were of analytical grade and used as received. Distilled water (18 MΩ cm) from a Millipore system was used in all studies. The sequences of the DNA strands used in the experiments are as follows: probe DNA: 5'-AGG-TCG-CCG-CCC-SH-3', target DNA: 3'-TCC-AGC-GGC-GGG-5', single-base mismatched DNA: 3'-TCC-AGC-GGC-GTG-5'

Synthesis of Graphene Oxide (GO)

GO was synthesised by a modified Hummers methods [27]. Three grams of graphite was added to 12 ml H_2SO_4 and kept at 80°C for 4.5 h. After sonication at room temperature, the solution was filtered using 200-nm porous filter and obtained the pre-oxidized graphite powder. To exfoliate the pre-oxidized graphite powder into monolayer graphene sheets, 2 g powder and 15 g $KMnO_4$ were added into 120 ml H_2SO_4 with stirring and an ice-water bath was adopted to ensure the temperature remain below 10°C. The mixture was stirred for 2 h under ice-water bath. About 250 ml distilled water and 20 ml H_2O_2 (30%) were added to dilute the solution at room temperature. The suspending solution was precipitated for 12 h and the upper supernatant was collected and centrifuged. Successively, the GO powders were washed with 10% HCl and distilled water three times. The obtained GO was dispersed in distilled water to get a stable brown solution.

Preparation and Characterization of rGO-Au Nanocomposites

Prior to the deposition of Au nanoparticles, the GO sheets were chemically reduced assisted by microwave irradiation (MWI). In a typical experiment of chemical reduction of GO, 3 μl of hydrazine and 0.1 g SDS were added to 20 ml of GO aqueous solution (0.5 mg/ml). After rigorous shaking, the mixture was heated with microwave irradiation (900 W, 2.40 GHz) for 30 s and obtained a stable rGO sheets dispersions. The reduced GO sheets were centrifuged and washed three times with distilled water to remove the hydrate residue, and dissolved into the SDS aqueous solution again. Au nanoparticles were deposited on rGO sheets by a chemical reduction of $HAuCl_4$ in aqueous solution. In a typical procedure, 0.1 ml aqueous of 0.05 M $HAuCl_4$ was added into 5 ml rGO SDS aqueous dispersion in a 10-ml bottle under rigorous stirring for 1 h. After the reduction reaction, the stable graphene–Au dispersion was obtained. The UV–Vis absorption curves of the resulting dispersions were measured using a UV–Vis spectrometer. The morphology of the rGO-Au nanocomposites was observed by transmission electron microscopy (TEM), tapping mode atomic force microscope (AFM), and scanning electron microscopy (SEM). Raman spectra were obtained with a WITec CRM200 confocal Raman microscopy system (laser wavelength 488 nm and laser spot size about 0.5 mm); the Si peak at 520 cm^{-1} was used as a reference for wavenumber calibration.

Fabrication and Characterization of rGO-Au FETs

The rGO or rGO-Au sheets were deposited by dip-coating method onto Si substrate with 300 nm thermal oxide on top first. To fabricate bottom-gated graphene FETs, selected monolayer rGO or rGO-Au sheet was located under a microscope, and then covered by a copper grid hard mask. The gold source and drain electrodes were subsequently deposited onto the sheets by thermal evaporation. The channel length between source and drain electrodes is 20 μm, and the thickness of gold electrode is about 50 nm. The electrical measurements were performed in ambient condition using a Keithley semiconductor parameter analyzer (model 4200-SCS). The effective carrier mobility was calculated using $\mu = (L/WC_{ox}V_d)(\Delta I_d/\Delta V_g)$ [28], where L and W are the channel length and width, C_{ox} the gate oxide capacitance, V_d the source–drain voltage, I_d the source–drain current, and V_g the gate voltage. The linear regime of the transfer characteristics was used to obtain the slope $\Delta I_d/\Delta V_g$.

Label-Free Electrical Detection of DNA Hybridization

It has been reported that the single-strand DNA with –SH end group can react with Au nanoparticles and can be used to realize DNA sensing [29]. To immobilize probe DNAs, the rGO-Au devices were immersed in

1 μM probe DNA solution for a period of 16 h. The samples were washed with PBS buffer to remove the weakly bound DNA, dried and characterized its electrical properties. Then, ten microliter target or mismatch DNA solution (200 nM) was pipetted onto the probe DNA decorated devices to hybridize 1 h, washed, dried, and characterized its electrical properties.

RESULTS AND DISCUSSION

Figure 1 shows the photographs of GO aqueous dispersion (0.5 mg/ml), rGO, and rGO-Au aqueous dispersion in presence of SDS. GO sheets form stable brown dispersion in water (Figure 1a), because the hydroxyl, epoxy, and carboxyl groups on the GO sheets render them hydrophilic [30]. However, the reduced GO sheets tend to form irreversible agglomerates and even restack to form graphite in water because of the π–π stacking interactions between rGO sheets. In order to obtain single- or few-layer graphene sheets, various methods have been attempted using polymers [31] or aromatic molecules [26] to prevent rGO aggregation. Here, we used SDS surfactant to obtain stable rGO aqueous dispersion (Figure 1b). However, precipitation occurs after a few days. After the modification of rGO with Au nanoparticles, the resulting rGO-Au is very stable (Figure 1c) (no precipitation at all in months). It is because the Au nanoparticles on the rGO sheets act as spacers to prevent the stacking interaction [32]. It is useful for the development of printable graphene nanoelectronics.

As shown in Figure 1d, the UV–Vis spectrum of the rGO dispersion in SDS solution (black line) possesses one absorption peak at ~271 nm, which is similar to rGO in water [33]. Also, there is a weak absorption peak at ~306 nm due to incomplete reduction of GO. In contrast, the adsorption spectrum of rGO-Au nanocomposite dispersion (red line) peaks at ~570 nm, indicating that the Au nanoparticles firmly attach to the surface of the rGO sheets. The reaction mechanism for Au nanoparticles formed (reduced) on the rGO sheets may be attributable to the difference of chemical potential between rGO and gold. This reaction mechanism has been proposed to explain the decoration of carbon nanotubes with metal nanoparticles [34].

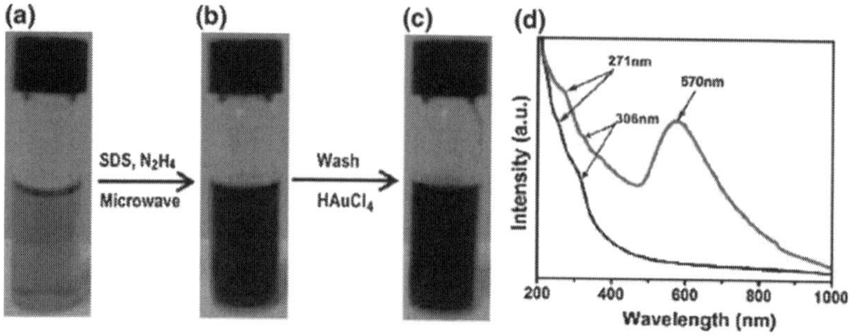

Figure 1. Photographs of ***(a)*** GO dispersion in water, ***(b)*** rGO dispersion in SDS solution, c rGO-Au nanocomposites in SDS solution. d UV–Vis absorptions of rGO (*black line*) and rGO-Au nanocomposites (*red line*) in SDS solution.

Single-layer rGO or rGO-Au sheets can be readily obtained using dip-coating of their dispersion. As revealed by AFM (Figure 2a), rGO sheet is smooth, with a thickness of ~1.10 nm which is somewhat larger than that of the mechanically exfoliated graphene monolayer (0.35 nm) due to the presence of epoxy groups on both sides of rGO sheet [35]. In contrast, numerous nanoparticles are observed on the rGO-Au sheet. As seen from Figure 2a, the Au nanoparticles (~10 nm in size) distribute uniformly on rGO sheet except a few large aggregates. TEM (Figure 2b) and SEM (Figure 2c) were also used to characterize the surface morphology of the rGO-Au sheet, where Au nanoparticles can be clearly identified. Interestingly, the density of Au nanoparticles is relatively higher at the edge of rGO sheet (Figure 2c) likely because of the higher surface energy (chemical potential) and more functional groups at the edge of rGO sheet as suggested previously [36].

Figure 3a shows the Raman spectra of rGO and rGO-Au nanocomposite. The Raman spectra both display two prominent peaks at ~1,350 and ~1,600 cm^{-1}, which correspond to *D* and *G* bands of graphene, respectively [37]. The *G* band corresponds to the E_{2g} mode and *D* band is attributed by in-plane A_{1g} zone-edge mode. When compared to the spectrum of rGO, the Raman *G* band slightly shifts from 1,602 to 1,606 cm^{-1} due to the *p*-doping effects imposed by the Au nanoparticles [18]. Furthermore, the 2D Raman map of rGO-Au constructed by integrating *G* band (Figure 3b) is similar to that of mechanically exfoliated single-layer graphene, indicating that our rGO-Au sheet is

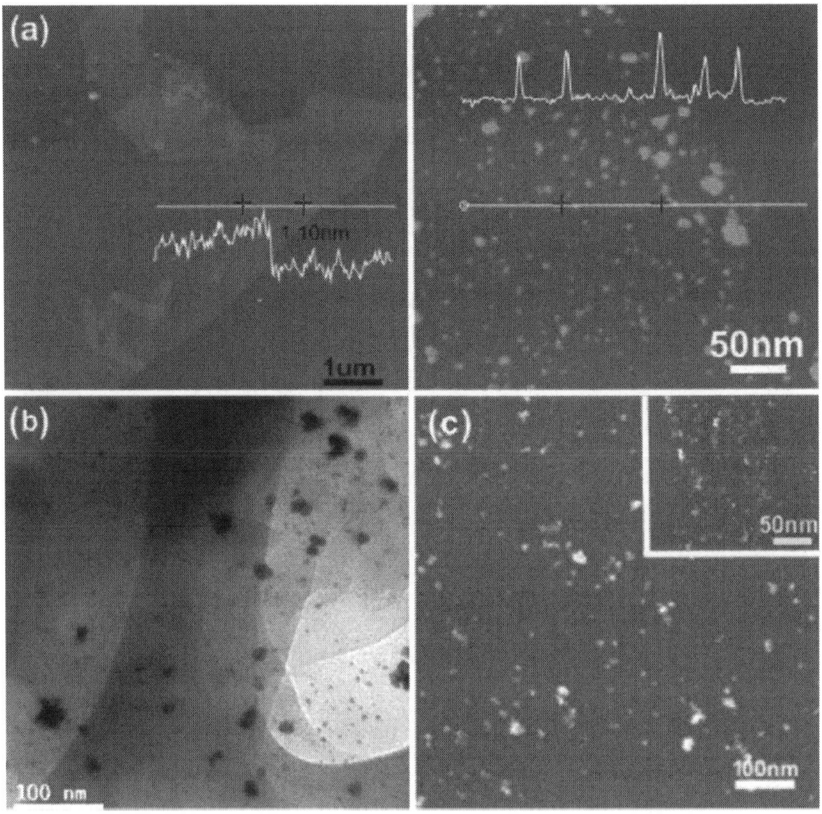

Figure 2. ***a*** AFM images of rGO sheets (***left***) and rGO sheets decorated with Au nanoparticles (***right***). The *insets* show the height profiles at the indicated places. *b* TEM image of rGO-Au sheet. ***c*** SEM image of rGO-Au sheet. The *inset* shows the view at the edge of rGO-Au sheet.

uniformly single-layered without significant defects. On the other hand, the 2D Raman map on the same area constructed by integrating 1,100 ~ 1,200 cm^{-1} region of Raman spectrum (Figure 3c) clearly indicates the location of Au nanoparticles (bright spots) because Au enhances Raman signal in this wavelength range [38]. Consistent with AFM, TEM and SEM imaging, these results show that Au nanoparticles are tightly attached to the surface of rGO sheets and uniformly distributed.

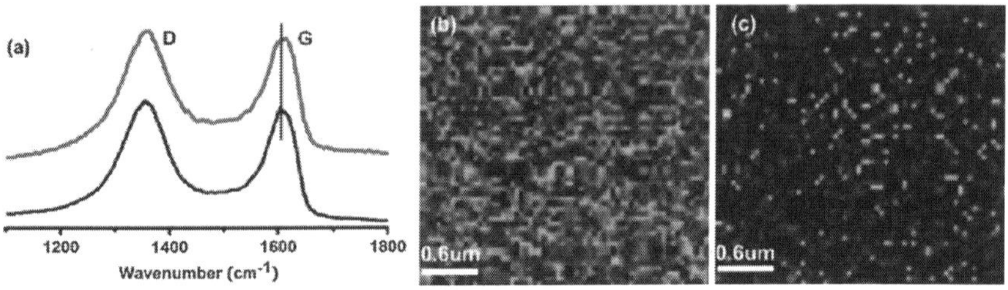

Figure 3. ***a*** **Raman spectra of rGO sheet (** ***black line*** **) and rGO-Au sheet (** ***red line*** **).** ***b*** Raman map of rGO-Au nanocomposite constructed by integrating 1,500–1,700 cm^{-1} region (G band) of graphene Raman spectrum. ***c*** Raman map of the same area constructed by integrating from 1,100 to 1,200 cm^{-1} in which the Au nanoparticles are indicated by the *bright dots.*

To investigate the effects of Au nanoparticles on the electrical properties of rGO sheets, the FETs with single-layer rGO or rGO-Au as the conducting channel were fabricated on SiO_2/Si substrates. The linear I_d–V_d curves of the devices indicate a good ohmic contact between the Au electrodes and the conducting channel (Figure 4). Both rGO FETs and rGO-Au FETs manifest typical V_g-dependent ambipolar field-effect characteristics, similar to pristine graphene [1], except that the Dirac (minimum current) point significantly shift to a more positive voltage likely due to *p*-doping from oxygen or moistures on rGO. The hole mobility of the rGO and rGO-Au FETs (slope of the I_d–V_g curve at *p*-region) is about 0.015 and 9.0 cm^2/Vs, respectively. This result indicates that the hole mobility of the rGO-Au sheet is several orders higher than that of rGO sheets (8.62 ± 2.54 cm^2/Vs vs. 0.03 ± 0.015 cm^2/Vs, $n = 8$ devices for each case). The enhanced mobility could be explained by reduction of defect sites on rGO by decorated Au nanoparticles, because the nucleation and growth of Au nanoparticles preferentially take place at the defect sites which have higher chemical potential and more reactive functionalities [39]. Therefore, the scattering effects caused by the defects are likely relived by nanoparticles.

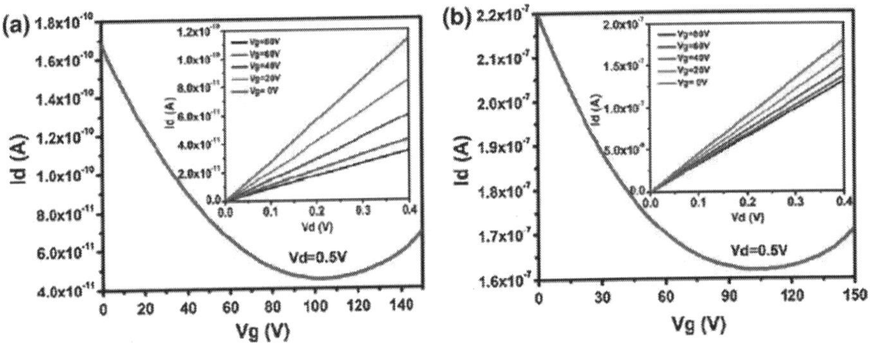

Figure 4. Typical $I_d - V_g$ ***curve*** of single-layer rGO-based FET (a) or rGO-Au-based FET (b). The *insets*show the I_d–V_d *curves* at different V_g.

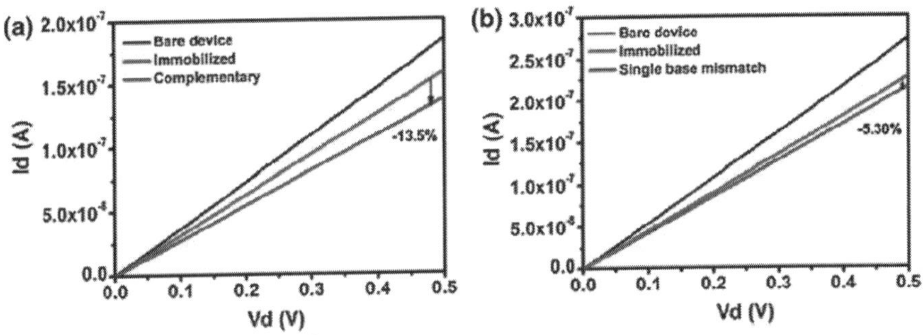

Figure 5. The drain current (I_d) versus drain voltage (V_d) curves (at V_g = 40 V) of the bare rGO-Au device, the same device with immobilized probe DNA, and the same device after hybridization of the complementary (a) or single-base mismatched (b) DNA strands (DNA concentration is 200 nM).

One of the promising applications of graphene electronics is for biosensing [13]. Here, we employed our rGO-Au FETs to detect DNA hybridization. Figure 5 shows the drain current (I_d) versus drain voltage (V_d) curves (at V_g = 40 V) of the bare rGO-Au device, the same device with immobilized probe DNA, and the same device after hybridization of the complementary or single-base mismatched DNA strands. As shown, the drain current decrease obviously after the immobilization of thiolated probe DNA on Au nanoparticles due to *n*-doping effect of DNA molecules. After the hybridization with the complementary target DNA, the drain current decreases further about 13.5%. In a parallel experiment, the target DNA strands with single-base mismatch only caused about 5.3% decrease in drain current after hybridizing with the probe DNA. The current decease caused by the complementary DNA is significantly larger than that caused by the single-base mismatched DNA (14.07 ± 0.58% vs. 6.02 ± 0.59%, $p < 0.001$, $n = 4$ devices for each case). Therefore, our rGO-Au FET sensors are able to discriminate DNA sequence with high sensitivity. Compared to the carbon nanotube network FETs, it has the similar detection sensitivity at the same DNA concentration [40].

CONCLUSIONS

In summary, we have demonstrated a simple approach to prepare stably dispersed rGO-Au sheets in aqueous media. Compared with the rGO sheet, the rGO-Au sheet FETs exhibit much higher hole mobility. In addition, the Au nanoparticles provide readily functionalization sites for conjugating biomolecular probes on the rGO-Au nanocomposites. We further showed that rGO-Au-based FETs are able to detect DNA hybridization with sequence specificity.

ACKNOWLEDGEMENTS

We acknowledge the financial support from the NSF of China (50902071, 61076067), the 973 Program (China, 2009CB930601), the NSF of the Education Committee of Jiangsu Province (TJ207035), the Science Foundation of Nanjing University of Posts and Telecommunications (NY208058), and A-Star SERC grant (No. 072 101 0020) of Singapore.

REFERENCES

1. Novoselov KS, Geim AK, Morozov SV, Jiang D, Zhang Y, Dubonos SV, Grigorieva IV, Firsov AA: *Science*. 2004, **306:** 666.
2. Zhang Y, Tan YW, Lstormer H, Kim P: *Nature*. 2005, **438:** 201.
3. Geim AK, Novoselov KS: *Nat Mater*. 2007, **6:** 183.

4. Jiao LY, Zhang L, Wang XR, Diankov G, Dai HJ: *Nature.* 2009, **458:** 877.

5. Geim AK: *Science.* 2009, **324:** 1530.

6. Stankovich S, Dikin DA, Dommett GHB, Kohlhaas KM, Zimney EJ, Stach EA, Piner RD, Nguyen ST, Ruoff RS: *Nature.* 2006, **442:** 282.

7. Bunch JS, Zande AM, Verbridge SS, Frank IW, Tanenbaum DM, Parpia JM, Craighed HG, McEuen PL:*Science.* 2007, **315:** 490.

8. Xu Y, Bai H, Lu G, Li C, Shi G: *J Am Chem Soc.* 2008, **130:** 5856.

9. Dong XC, Shi YM, Zhao Y, Chen DM, Ye J, Yao YG, Gao F, Ni ZH, Yu T, Shen ZX, Huang YX, Chen P, Li LJ: *Phys Rev Lett.* 2009, **102:** 135501.

10. Gilje S, Song H, Wang M, Wang KL, Kaner RB: *Nano Lett.* 2007, **7:** 3394.

11. Stampfer C, Schurtenberger E, Molitor F, Guttinger J, Ihn T, Ensslin K: *Nano Lett.* 2008, **8:** 2378.

12. Schedin F, Geim AK, Morozov SV, Hill EW, Blake P, Katsnelson MI, Novoselov KS: *Nat Mater.* 2007, **6:** 652.

13. Huang YX, Dong XC, Shi YM, Li CM, Li LJ, Chen P: *Nanoscale.* 2010, **2:** 1485.

14. Kim YK, Min DH: *Langmuir.* 2009, **25:** 11302.

15. Xu C, Wang X, Zhu JW: *J Phys Chem C.* 2008, **112:** 19841.

16. Wang S, Tambraparni M, Qiu JJ, Tipton J, Dean D: *Macromolecule.* 2009, **42:** 5251.

17. Agarwal S, Zhou XZ, Ye F, He QY, Chen G, Soo JC, Boey F, Zhang H, Chen P: *Langmuir.* 2010, **26:** 2244.

18. Dong XC, Fu DL, Fang WJ, Shi YM, Chen P, Li LJ: *Small.* 2009, **5:** 1422.

19. Park S, Ruoff RS: *Nat Nanotechnol.* 2009, **4:** 217.

20. Robinson J, Weng XJ, Trumbull K, Cavalero R, Wetherington M, Frantz E, LaBella M, Hughes Z, Fanton M, Snyder D: *ACS Nano.* 2009, **4:** 153.

21. Reina A, Jia XT, Ho J, Nezich D, Son H, Bulovic V, Dresselhaus MS, Kong J: *Nano Lett.* 2008, **9:** 30.

22. Wang HL, Robinson JT, Li XL, Dai HJ: *J Am Chem Soc.* 2009, **131:** 9910.

23. Su CY, Xu YP, Zhang WJ, Zhao JW, Tang XH, Tsai CH, Li LJ: *Chem Mater.* 2009, **21:** 5674.

24. Gao XF, Jang J, Nagase S: *J Phys Chem C.* 2009, **114:** 832.

25. Pasricha R, Gupta S, Srivastava AK: *Small.* 2009, **5:** 2253.

26. Huang YX, Chen P: *Adv Mat.* 2010, **22:** 2818.

27. Hummers WS, Offeman RE: *J Am Chem Soc.* 1958, **80:** 1339.

28. Dong XC, Su CY, Zhang WJ, Zhao JW, Ling QD, Huang W, Chen P, Li LJ: *Phys Chem Chem Phys.* 2010,**12:** 2164.

29. Dong XC, Lau CM, Lohani A, Mhaisalkar SG, Kasim J, Shen ZX, Ho XN, Rogers JA, Li LJ: *Adv Mater.* 2008, **20:** 2389.

30. Stankovich S, Piner RD, Chen XQ, Wu NQ, Nguyen ST, Ruoff RS: *J Mater Chem.* 2005, **16:** 155.

31. Yu DS, Dai LM: *J Phys Chem Lett.* 2009, **1:** 467.

32. Hassan HMA, Abdelsaved V, Khder AR, AbouZeid KM, Terner J, El-Shall MS, El-Azhary A: *J Mater Chem.* 2009, **19:** 3832.

33. Li D, Muller MB, Je SG, Kaner RB, Wallace GG: *Nat Nanotechnol.* 2008, **3:** 101.

34. Choi HC, Shim M, Bangsaruntip S, Dai HJ: *J Am Chem Soc.* 2002, **124:** 9058.

35. Stankovich S, Dikin DA, Piner RD, Kohlhaas KA, Kleinhammes A, Jia YY, Wu Y, Nguyen ST, Ruoff RS:*Carbon.* 2007, **45:** 1558.

36. Zhou XZ, Huang X, Qi XY, Wu SX, Xue C, Boey FY, Yan QY, Chen P, Zhang H: *J Phys Chem C.* 2009, **113:**10842.

37. Kudin KN, Ozbas B, Schniepp HC, Aksay IA, Car R: *Nano Letters.* 2008, **8:** 36.

38. Cao YC, Jin R, Mirkin CA: *Science.* 2002, **297:** 1536.

39. Goncalves G, Marques PA, Granadeiro CM, Nogueira HS, Singh MK, Gracio J: *Chem Mater.* 2009, **21:**4796.

40. Dong XC, Fu DL, Xu YP, Wei JQ, Shi YM, Chen P, Li LJ: *J Phys Chem C.* 2008, **112:** 9891.

11

CARBON NANOTUBES AND GRAPHENE NANORIBBONS: POTENTIALS FOR NANOSCALE ELECTRICAL INTERCONNECTS

Caterina Soldano [1,*], Saikat Talapatra [2] and Swastik Kar [3]

[1]ETC srl, Bologna 40129, Italy

[2]Physics Department, Southern Illinois University at Carbondale, Carbondale, IL 62901, USA

[3]Department of Physics, Northeastern University, Boston, MA 02115, USA

ABSTRACT

Carbon allotropes have generated much interest among different scientific communities due to their peculiar properties and potential applications in a variety of fields. Carbon nanotubes and more recently graphene have shown very interesting electrical properties along with the possibility of being grown and/or deposited at a desired location. In this Review, we will focus our attention on carbon-based nanostructures (in particular, carbon nanotubes and graphene nanoribbons) which could play an important role in the technological quest to replace copper/low-k for interconnect applications. We will provide the reader with a number of possible architectures, including single-wall as well as multi-wall carbon nanotubes, arranged in horizontal and vertical arrays, regarded as individual objects as well as bundles. Modification of their functional properties in order to fulfill interconnect applications requirements are also presented. Then, in the second part of the Review, recently discovered graphene and in particular graphene and few-graphene layers nanoribbons are introduced. Different architectures involving nanostructured carbon are presented and discussed in light of interconnect application in terms of length, chirality, edge configuration and more.

Keywords: carbon nanotubes; graphene nanoribbons; nanoscale interconnects

1. INTRODUCTION

Over the past few decades, the increasing progress in most technological areas from biomedicine to strategic deployment to meteorological forecasting has been largely benefitting from continuous and massive advances achieved in the field of computational technologies. With the need for larger quantities of data processing at faster pace within smaller volumes, the microelectronics industry has so far been steadily downsizing its basic active and passive building blocks. Those components are currently capable of operating at higher clock-speeds, delivering higher number of computations per second and, at the same time they occupy less physical space and consume less power. Nevertheless, the rapid progress towards gigascale miniaturization of those components faces a number of difficult challenges in terms of materials, architectures, fabrication and integration of nanoscale active and passive elements with expected performance in terms of reliability, speed, compatibility, and power consumption [1]. In this scenario, a fundamental issue of considerable difficulty is the development of the next generation interconnects for microprocessor units (MPUs). Interconnects are wiring systems which distribute clock and other signals and provide power and ground, to and among the various circuits and systems functioning on a chip. As the various active and passive elements of the MPU reduce in dimension below 100 nm, also the corresponding interconnects will consistently follow. Existing interconnect technologies involving Cu/low-k lines and theirs evolutionary downsizing for sub-100 nm size interconnects face a number of important challenges. As the lateral dimension of interconnects approaches the mean free path of copper (~40 nm at room temperature [2]), the impact of grain boundary scattering, surface scattering, and the presence of a high-resistivity material as a diffusive barrier layer causes a rapid increase in the overall resistivity. Figure 1a shows how the resistivity of copper significantly and rapidly increases above its bulk value as interconnect lateral dimension shrinks below 100 nm [3], following which major degradation of reliability, signal transfer properties and thermal management issues start to become important. In addition, larger current densities for increasingly small interconnect lines will lead to an enhanced electro-migration failure. Tightly packed conducting material immersed in a dielectric environment will further introduce significantly larger parasitic capacitance, enhancing the delay of the signal propagation and limiting the overall operational bandwidth. This scenario clearly suggests that any length below 22 nm, which represents the current microelectronics community standard, copper will reveal to be inadequate for interconnect applications. Hence, a revolutionary new high-conducting material with higher failure current density and comparable or better performance metrics (such as specific capacitance and RC delay) is indeed in need for future copper replacement. Figure 1b represents the cross-sectional schematic of a typical Integrated Circuit (IC), reproduced from the International Technology Roadmap for Semiconductors (ITRS) map from 2009. In a typical IC, interconnect structures are scaled into local, intermediate, and global wiring levels based on optimization of delay, power, and bandwidth *etc.* and where each level has diverse requirements and at the same time poses different challenges. At the local (or M1) levels, resistivity and failure current density represent major bottlenecks, mostly in terms of delay. Global interconnects, which are the longest, will be impacted most by the degraded RC delay (*i.e.*, interconnect RC delay of 1892 ps for 1 mm Cu intermediate wire is thus not compatible). Significant effort has also been invested in tuning and modifying the dielectric properties of the interconnect material itself. Although the effective dielectric constant has been predicted to be progressively reduced (from 4 to 1.5) with developing technologies [4], nevertheless we have now reached a point in which no further important decrease is possible. Currently, the lowest value of the effective dielectric constant ranges between 2.5 and 3 and it has been stable over the last few years [5]. While some progress has been made in terms of evolutionary downsizing of existing technologies which will possibly work in the near-term (~22 nm), it is becoming increasingly evident that revolutionary new materials and/or processes are required to address long-term (<16 nm) scaling down of interconnects.

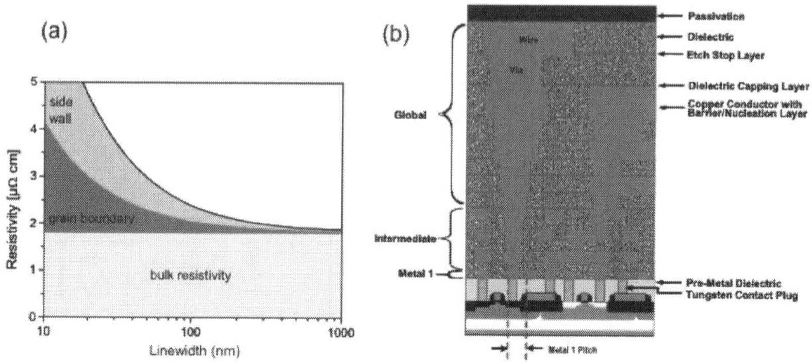

Figure 1. (**a**) Resistivity of copper as function of the interconnect linewidth; (**b**) Typical cross-section of hierarchical scaling (microprocessor unit (MPU) device) showing different interconnect scaling levels [3].

Different replacements for copper-based architectures have been proposed in the form of 3D Integrated Circuits, optical interconnects and carbon nanotubes (CNTs). Ongoing research in each of these field shows that those materials and structures could possibly successfully overcome the delay, power and bandwidth limitations of traditional interconnects, at one or more scaling levels. Nonetheless, in every single case, much remains to be done before one or more of these technologies can be realistically integrated with existing CMOS platforms while retaining their novel high-performance capabilities.

In this Review, we will focus our attention on carbon-based nanostructures (in particular carbon nanotubes and graphene nanoribbons) which could play an important role in the technological quest to replace copper/low-k for interconnect applications. We will provide the Reader with a number of possible architectures, including single-wall as well as multi-wall carbon nanotubes, arranged in horizontal and vertical arrays, and regarded as individual objects as well as bundles or arrays. Modification of their functional properties in order to fulfill interconnect applications requirements are also presented. Then, in the second part of the Review, recently discovered graphene and in particular graphene and few-graphene layers nanoribbons are introduced. Different architectures involving nanostructured carbon are presented and discussed in light of interconnect applications in terms of length, chirality, edge configuration and more.

We refer the Reader to more detailed literature regarding alternatives to carbon-based interconnects here presented (3D Integrated Circuit and Optical Interconnects) [6,7,8].

2. CARBON NANOTUBES

The ability to withstand current densities larger than 10^9 A/cm^2 (more than three orders of magnitude more than copper) even at ~1 nm diameter distinctly renders carbon nanotubes (CNTs) ideal candidates for nanoscale electrical interconnects [9]. CNTs are one-dimensional (1D) quantum conductors that have been shown to have ballistic transport with mean free paths of the order of microns [10,11], which is significantly larger than copper. Carbon nanotubes can have diameters ranging from <1 nm (Single-Wall Nanotubes, SWNTs) up to tens of nanometers (Multi-Wall Nanotubes, MWNTs). While SWNTs can be either metallic or semiconducting, MWNTs are mostly metallic. In addition to their attractive ballistic conduction capabilities, these chemically and mechanically stable strands of carbon are extremely strong (tensile strength ~ 100 GPa) and possess very high thermal conductivity [12,13]. It has been shown that metallic SWNTs present values of the conductance close to their theoretically predicted values of 4 e^2/h and they are capable of functioning at tens of GHz range both as transistors and as interconnects [14,15]. Above everything else, CNTs have the natural geometrical shape for being a wire, *i.e.*, a nanometers-diameter cylinder with a large aspect ratio (it will be shown later on that it is easy to fabricate SWNTs that are microns long, and MWNTs that are centimeters long). All those favorable physical, electrical and geometric properties of carbon nanotubes make them an attractive and suitable choice as nanoscale electrical interconnects [16,17].

As compared to Cu/low-k interconnects, CNTs offer numerous advantages, as a results of the very strong nature of the C-C bonds, their 1D tubular shape and the electronic band structure. To benchmark Cu-based interconnects, carbon nanotubes have specific properties and fulfill the following requirements: (i) *reduced resistance*, in fact in the limit of short interconnects there exists a limitation associated to the quantum resistance of a single tube (for metallic SWNTs, the quantum of resistance is 6.5 kΩ [18], while for long interconnects the conductivity of nanotubes in densely-packed bundles is predicted to outperform copper counterpart; (ii) *resistance to electromigration*, due to the strong carbon-to-carbon bonds within the hexagonal lattice of the shell(s), resulting in the possibility by the tube to sustain elevated current density (10^9 A/cm^2), while for copper equivalent current density of 10^6 A/cm^2 already is capable of producing structural damages; (iii) large *thermal conductivity*, in fact isolated carbon nanotubes are predicted to have large values of thermal conductivity (few thousands W/mK [19,20]).

Increased resistance resulting when reducing size in *vias* structures can be addressed using bundles of carbon nanotubes; an additional degree of improvement can be obtained by increasing the site density. Further, since the energy gap of carbon nanotubes is inversely proportional to their diameters, it is important to control the growth of CNTs, which is always a mixture of metallic as well as semiconducting SWNTs, thus small-diameter SWNT must be reduced since will limit the conduction of the bundles and *vias*architectures.

2.1. Carbon Nanotubes for High-Density Integration

In nearly all calculations and most experiments, carbon nanotubes in the form of aligned arrays have been used to benchmark future interconnects against copper existing architectures. These structures could be of circular or rectangular cross sections, made of SWNTs or MWNTs bundles, and can be commonly treated as parallel nanotube architectures (PNAs). Those structures can potentially replace Cu/low-k interconnects at most levels of interconnect hierarchy except in places where low-resistance short interconnects are needed (*i.e.*, local interconnects), where CNT interconnects would be significantly more resistive than minimum-size copper wires since nanotubes as short as the gate pitch would be needed. In addition to the *classical*problems such as scattering associated with downsizing of copper interconnects, at the nanoscale, quantum confinement effects additionally play an important role. Recent computational predictions [21] have demonstrated that from a purely quantum-mechanical point of view, the resistance of carbon nanotube bundles is smaller than that of the copper wires for dimensions below 60 nm. Further, transmission line models [22] suggest that while ohmically contacted all-metallic SWNT bundle interconnects at local levels may be equivalent to copper in terms of RC delay, nevertheless the reduction in power dissipation and the increase of current density rendered them preferable even for the local level case. At the intermediate or global levels, RC delay becomes a crucial factor. In this case, larger diameter MWNT bundles, being chirality-independent all-metallic and having large mean free-paths, can achieve smaller signal delay than copper interconnects, with the improvements becoming more significant with technology scaling and increasing wire lengths. Saraswat *et al.* have predicted that for global levels, carbon nanotubes PNAs will outperform copper both in terms of latency as well as energy per bit of operation [23]. Monolayer or multilayer SWNTs PNAs can offer up to 50% reduction in capacitance and power dissipation with up to 20% improvement in latency if short enough (less than 20 μm) [24,25,26,27]. For semi-global interconnects, both latency and power dissipation can be substantially improved using bundles of SWNTs with reduced cross-sectional dimensions. From the electrical point of view, MWNT PNAs can outperform both SWNT and Cu at intermediate and global length scale in terms of conductivity. In addition, all-metallic SWNT PNAs appear to have much lower capacitance per unit length compared to copper and hence are ideal for local interconnects in terms of latency and power dissipation. This, in the long-term (sub-22 nm technology node) at all scaling levels, densely packed, all metallic PNAs, either SWNTs or MWNTs would be suitable copper replacements in terms of resistivity, failure current density, power dissipation, and delay. In the following sections, we will discuss some recent technological breakthroughs in terms of materials development, processing, structure fabrication, performance characterization, performance enhancement, and modeling of realistic SWNT and MWNT architectures that can potentially overcome the challenges faced by downward scaling of Cu/ low-k interconnects in terms of their projected performance metrics.

2.1.1. Scalable Ultra-High Density, Super Aligned SWNT PNAs

In order to compare or outperform Cu in terms of conductivity, it is essential to build highly aligned, short or/and long parallel arrays of CNTs with very high number density. For 1 nm diameter SWNTs the ideal density for a densely-packed all-metallic bundle is 0.66 SWNTs/nm^2 has been proposed [1]. Jung *et al.* [28] have developed a template-based fluidic assembly process to build highly organized aligned single-wall nanotubes PNAs on SiO_2/Si substrate. Figure 2a schematically describes the steps involved in the fabrication of these structures. First, a plasma treatment is used to enhance the hydrophilic nature of SiO_2 surface, on which subsequently lithographically patterned photoresist is used to fabricate "channels". Substrates are then dip-coated into an appropriately functionalized

SWNTs dispersion and the tubes selectively adhere to the SiO_2 trenches. Removal of the resist leaves behind highly organized, dense, aligned arrays of SWNTs, the thickness of which (~tens of nanometers) can be controlled by the initial SWNT concentration and dipping speed. The dense filling factor and high degree of alignment of the structure is maintained end-to-end of the PNA structure, as shown in Figure 2b. The lateral size of the interconnect structures is limited only by the lithography process, with which sub-100 nm channels can be easily achieved. So-fabricated devices are characterized by an overall resistance ranging from few kΩ to tens of kΩ depending on the channel size (seeFigure 2c), and they are able to withstand current density of $\sim 10^7$ A/cm^2. A very important and encouraging aspect of this technique is that the degree of alignment and the overall density appears to increase with smaller channel widths, which represents a great advantage in the perspective of interconnect applications. Further, the possibility of increasing the length in case of SWNTs by adding titanium to the catalyst surface leads to millimeter-long, small-diameter, single-wall carbon nanotube as a results of enhanced oxygen affinity [29]. Additional approach have also been proposed to fabricate SWNTs two-dimensional arrays such as electric field-assisted direct growth [30], low-feeding gas flow driven method [31] and AC dielectrophoresis (DEP) [32] from a solution of SWNTs, the last one leading to planar arrays of controllable density between pre-fabricated electrodes ranging from 0.5 SWNT/μm to more than 30 SWNT/μm, depending on the concentration of tubes in solution, as shown in Figure 2d,e. This approach leads to highly oriented PNAs with sheet resistance as low as few kΩ/□ depending on the density of carbon nanotubes (much lower compared to thin film of random nanotubes or nanotube cross bar junctions [33]) and the electrical properties of which can be tuned by the removal by breakdown of metallic SWNTs. Han and co-workers have demonstrated that the modification of the surface properties of single wall carbon nanotubes with short DNA linkers can favor the formation of parallel arrays where nanotubes are arranged one parallel to the others. In this case, DNA behaves as a spacer with a potential control of the inter-distance between adjacent tubes of less than 3 nm and the entire arrays can be easily transferred [34]. In principle this approach is very powerful (not limited to carbon nanotubes), however a post-fabrication densification step is in need to further increase the density of tubes.

2.1.2. Densely Packed Ultra-Long MWNTs

Multi-wall carbon nanotubes PNAs are ideally suitable for intermediate or global level interconnects since they are nearly all metallic due to larger diameters and they are characterized by μm-long mean free path. Each shell of properly contacted MWNTs can contribute to the conduction leading to low resistance conductive wires. Long MWNTs, with lengths ranging from 500 μm (intermediate) to 1000 μm (global), could potentially achieve significantly smaller signal delays [35] (as low as 15%) compared to copper equivalent. Various methods have been developed to grow very long and aligned MWNT bundles, with high yield and repeatability [36,37,38]. Li *et al.* have shown that ultra-long bundles can be grown in highly aligned parallel arrays of MWNTs, each nanotube running continuously end-to-end and that a selective control over the tube diameter (usually in the 25–40 nm range) can be obtained by adjusting the growth parameters [39]. Extraordinary growth can also be achieved through a multi-step *bottom-up* growth of self-supporting vertically aligned MWNTs, which can easily reach mm long *Vias*-type structures with large degree of alignment [40], as shown in Figure 3.

In addition, closely-packed carbon nanotube bundles are predicted to be characterized by higher conductivity when compared to copper; however, it is still extremely difficult to controllably grow those structures. Varying the feedstock flux during the CVD process (also known as "pulsed" CVD) can lead to dense array of all vertically aligned nanotubes: changing the gas flux and the duration of the pulse allows for the growth in predetermined length intervals, with precise density of the array over sub-μm lengths and tailored heights if growth can be reinitiated after each pulse [41]. Design and engineering of the catalyst offer the possibility of growing ultrahigh density, aligned forests of carbon nanotubes. Cyclic and repeated deposition/annealing/immobilization of catalyst thin films lead to nanotube forests with an area density of at least $10^{13}/cm^2$, over 1 order of magnitude higher than existing values, and very close to the limit of a fully dense forest [42]. Additional improvement (up to $10^{14}/cm^2$) can be reached if tube diameter can be further reduced.

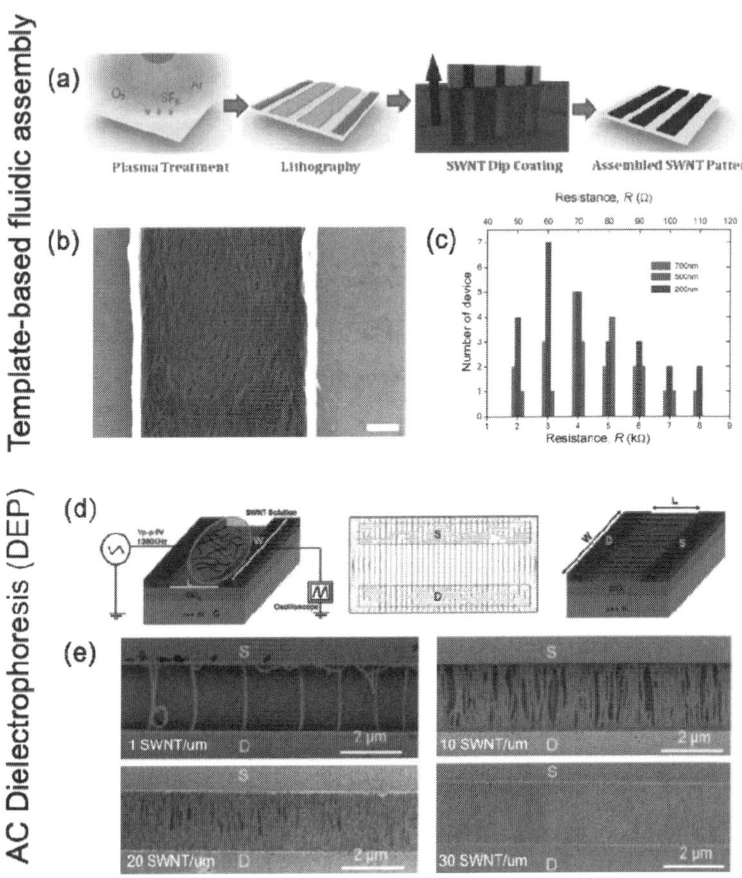

Figure 2. (**a**) Schematic of template-based fluidic assembly of high density, super aligned, sub-µm single-wall nanotube (SWNT) parallel nanotube architectures (PNAs); (**b**) High-magnification SEM images showing the degree of alignment of the SWNTs within channel widths of 1 µm. Scale bars = 200 nm; (**c**) Resistance histograms of SWNT-PNAs for different channel length (700, 500 and 200 nm, as labeled); (**d**) Schematic of the dielectrophoresis (DEP) assembly method; (**e**) SEM images for a typical DEP assembly with varying nanotube densities (1, 10, 20 and 30 tubes/µm). [28,32]

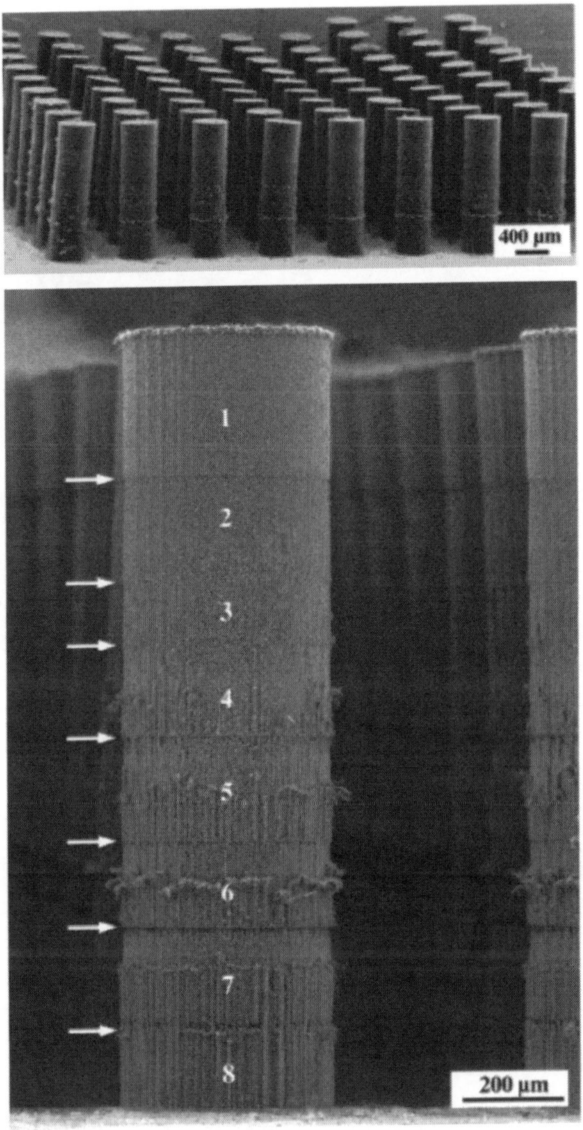

Figure 3. (Top) SEM image showing a two-dimensional array of pillars, each made of eight stacks of vertically aligned nanotube layers. Eight separate chemical vapor deposition (CVD) steps were used to grow the eight stack pillars on pre-patterned SiO_2. (Bottom) Higher-magnification SEM image showing the interfaces indicated by arrows between the separate stacks of nanotubes in a single pillar. The stacks are numbered 1 to 8, corresponding to the first to the eighth stack grown sequentially [40].

Post-growth methods have also been proposed in order to densify MWNTs arrays. Natural shrinking has been observed after immersion in solution due to simple solvent evaporation [43]. In as-grown aligned CNT arrays, the site density of nanotubes is about 60 CNTs/μm^2, orders of magnitude smaller than the effective maximum achievable site densities (~10^4 CNTs/μm^2). A solvent-assisted capillary coalescence method [44] is used to densify aligned CNT arrays in vertical pillars, shape very well suitable for vertical interconnect assemblies and vias, as shown in Figure 4.

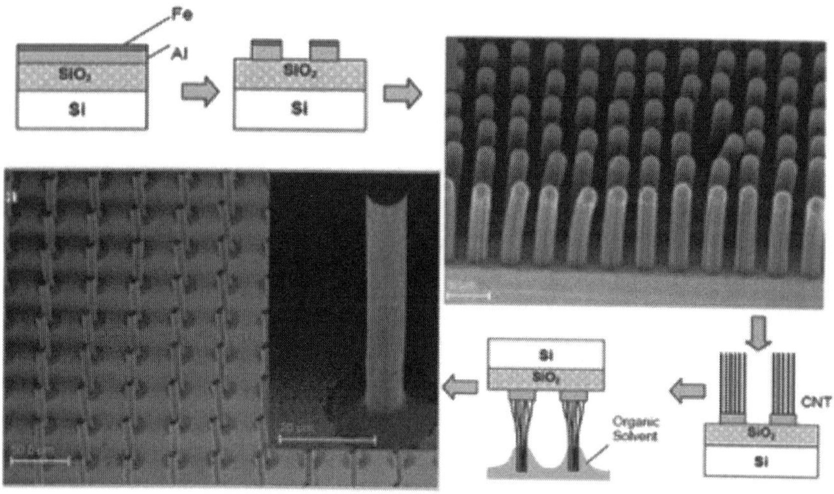

Figure 4. Schematics and SEM images describing capillary coalescence-driven densification using an organic solvent in aligned MWNT PNAs [44].

The resultant pillars retain their cylindrical nature while uniformly shrinking in diameter, with site densities of approximately 900 CNTs/μm^2, with a densification factor of about 15. However, such structures will still present a significant unoccupied volume between individual CNT strands, and this could provide room for further densification. On the other hand, growth in templates (alumina, in most cases) allows for selective fabrication of all-vertically aligned and closely packed tubes with diameter as small as 30 nm [45]. Those structures, when still in template, could in principle allow for *all-walls top and bottom electrically contacted* nanotubes with uniform length and diameter. However, tubes grown in template are of poor quality when compared to arc-discharge tubes with a high level of defects which makes them highly resistive and, at the same time, the density (tube/μm^2) is very low.

2.1.3. Selective Growth of Metallic Tubes

Statistically, about one third of as-grown SWNTs are metallic. These single shell tubes realistically represent the ultimate interconnects due their small radius and high conductance. However, there are no simple methods for selectively grow only metallic nanotubes, especially in more complex structures. Although great efforts have been put toward growth of controlled-diameter [46,47] and metallic SWNTs, very few reports have shown a certain degree of control over the growth. Yield of metallic SWNTs up to 65% of the final product has been achieved using as carbon source during the growth process monohydroxy alcohol homologues with a different ratio of carbon to oxygen content. It has been demonstrated that the yield of metallic SWNTs can be improved by increasing the relative concentration of carbon to oxygen [48]. Further, Harutyunyan *et al.* were successful in achieving a fraction of metallic SWNTs of 91% by modifying the noble gas ambient during thermal annealing of the catalyst and in combination with oxidative and reductive species of the Fe nanocatalysts deposited on SiO_2/Si substrates. This work has clearly demonstrated the very strict connection between the catalysts morphology and the growing tube [49]. Modification of the catalyst properties (metallic phase and crystalline structure) can be a powerful and potential route in order to selectively tune the growth of carbon nanotubes with designated chirality [50,51]. Further on, detailed investigation of the effect of introducing iron pentacarbonyl ($Fe(CO)_5$) vapor during the growth process (arc-discharge) has shown the possibility of obtaining up to 90% of single wall metallic tubes [52]. Otherwise, it is possible to controllably grow double wall carbon nanotubes (DWNTs); in fact, due to their larger diameter, DWNTs have extremely small band-gaps, and at room or elevated temperatures, these tubes behave like near-zero gap metallic CNTs. Fu and coworkers have recently demonstrated the possibility of grow pillars on gold films of highly vertically aligned nanotubes, 80% of which are double wall carbon nanotubes [53], as shown in Figure 5. As compared to standard growth on Si substrate, it has been possible to considerably increase the number of DWNTs and less dispersive for other types of tubes. Water-assisted chemical vapor deposition on precisely-controlled thickness of iron catalytic films can be also applied to grow vertically aligned doubled-walled carbon nanotube forests with more than 2 mm in height with a yield of DWNTs approximately of 85% [54]. This synthesis approach also produces nearly catalyst-free DWNT forests with very high carbon purity (99.95%), which then potentially guarantees that the tube properties will not be affected by the presence of the catalyst. Koshio and coworkers have shown the possibility to obtain double-wall carbon nanotubes starting from highly dense small diameter MWNTs as a result of vacuum heat treatment at 2200 °C or higher, where, at these elevated temperature a selective removal of 1 to 3 shells within the MWNT takes place [55].

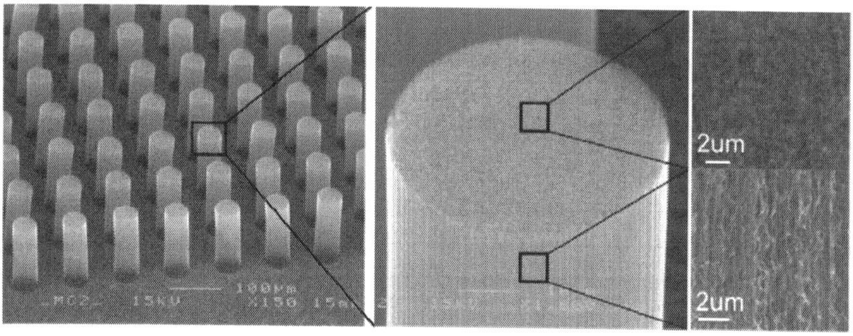

Figure 5. CNT bundles grown on the Au film, 80% of which are double wall carbon nanotubes (DWNTs). Top and side view show the large degree of vertical alignment of the tubes within the array [53].

Different post-growth approaches have also been proposed to selectively differentiate SWNTs, according to their chirality (and thus their electrical properties). An electric field-based treatment [56,57] can be used to separate all metallic tubes from semiconducting ones, leaving behind a parallel device array composed of a randomly grown mixture of metallic and semiconducting to reach densely packed devices in parallel where high density are required. In solution, single wall tubes can be well separated by density gradient ultracentrifugation (DGU), a technique traditionally used to separate and isolate different sub-cellular components, such as DNA and RNA. This method has been developed for sorting tubes based on chiral numbers and diameters, consisting in adding the SWNTs suspension in centrifuge tubes with varying density profile liquid mixtures; upon centrifugation, the tubes are separated and moved to regions (liquid) with matching densities [58,59,60]. Single-wall nanotubes can additionally be sorted using alternating current as a result of different dielectric constants depending on their electrical nature (metallic or semiconducting), as shown by Krupke and co-workers which have successfully separate in suspension metallic from semiconducting tubes [61].

2.1.4. Directional Growth of CNTs

Horizontal carbon nanotubes arrays represent still a challenge in the race for controllable and selective growth of carbon nanotubes. Interconnects applications are intrinsically in need of all the CNT to be of metallic nature. Being the energy gap of CNTs inversely proportional to the their diameter, the possibility of controlling the diameter during growth can allow for a control of the electronic properties of the tubes (and in particular their resistance [25,62,63]. There are three different approaches to achieve horizontal growth of carbon nanotubes: (i) electric-field based assembly/growth [64]; (ii) laminar gas flow [65] and (iii) step-templated growth on quartz/sapphire substrates [66,67]. Each of these methods unfortunately lead to very low nanotube densities or, as in the case of growth on quartz and sapphire, those are not compatible with growth on a chip. Further, one can think of first growing CNTs along the vertical direction and then proceed to flip them over.

Chai and coworkers have demonstrated that by controlling the catalyst thickness [68] on substrate (and the size of nanoparticles), it is possible to grow in the presence of an electric field bundles of MWNT with precise diameters along the horizontal direction [69]; in particular thicker catalyst films (5 nm) lead to large diameter tubes (25–35 nm) while in the case of thinner catalysts (1.2 nm), the diameter distribution of the grown tubes is around 5–7 nm. One other interesting outcome of this work is that catalyst thickness also allows for a certain control over the sites density of the CNTs bundles, where thinner catalyst leads to denser bundles.

Chiodarelli and coworkers have recently proposed a novel method to achieve horizontal growth of CNTs-based interconnects with lateral dimension comparable with current Cu equivalent [70]. CNTs are first grown vertically within pre-determined vias structures (diameters 200–300nm), then flipped horizontally and electrically contacted on both sides through metallization process. The as-fabricated interconnect structures then consist of carbon nanotube bundles with wall density of approximately $10^{13}/cm^2$, wire lengths of tens of μm and tube diameter of approximately 50 nm, characterized by nanotube resistivity as low as 1.1 mΩ·cm, value which is still larger than copper equivalent but which becomes closer and closer.

The placement of catalysts on a vertical surface makes the horizontal growth harder to achieve than the vertical growth. Yan and coworkers have shown that a tilt evaporation technique can achieve catalyst patterning on vertical surfaces [71]. Upon demonstrating that $CoSi_2$ is a good catalyst support material in CNT interconnect application since it can be easily patterned on a vertical surface and support high density CNT growth, they have grown CNTs from the vertical surface with wall density of 6×10^{11} walls/cm^2 [72]. In case of single-wall and double-wall tubes, this challenge can be achieved using for example previously described template-based fluidic assembly methods, in case of MWNTs this can be achieved, to a large extent, by using floating vapor catalyst assisted CNT growth directly on metals [73]. Talapatra *et al.* have developed a standardized and unique technique for growing 3-D architectures of aligned bundles of carbon nanotubes on bulk metals, as shown in Figure 6 [74].

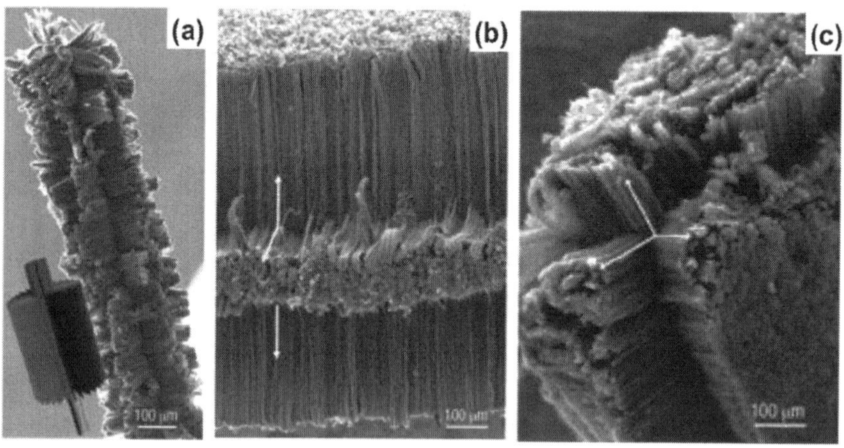

Figure 6. SEM of aligned MWNT arrays grown on metallic substrate showing growth directions conformal to the substrate orientation. Growth of carbon nanotubes on (**a**) metal wire, (**b,c**) metal sheets showing the highly directional nature of the CNT growth, with the orientation of the nanotubes being perpendicular to the substrate surface. The arrows indicate the direction of CNT growth [74].

Growth on metal substrates has also the advantage of additionally providing an easy solution for growing electrically end-contacted carbon nanotubes. Parameters such as growth temperature and growth time can be selectively adjusted in order to obtain a specific tube length, as well in part their degree of alignment [75]. Electrically end-contacted dense bundles of aligned CNTs are promising candidates for interconnects and *via*structures at the sub-50-nm length scale. For optimal performance in these architectures, any metallization of individual tubes has to be on their ends, such that individual shell can be accessed by each electrode, and each nanotube in the bundle has to be individually metalized in this way. Post-growth metallization techniques are yet to reliably meet these criteria on a large scale. The unique advantage of the direct growth of CNTs on metallic substrates is that, within the array, every individual nanotube is electrically end-connected *in-situ*, overcoming several obstacles to applications in interconnect and contact technologies.

2.2. Carbon Nanotubes Devices and Architectures

2.2.1. Towards Low-Resistance Contacts

Large contact resistance represents still a technological challenge in order to realize interconnects. In the case of single-wall nanotubes PNAs, the sidewall resistance can be dramatically reduced using a cyclic high-current induced contact resistance annealing. This causes "hot-spots" at the contacts since that is where the resistance lies. The current density ($\sim 10^8$–10^9 A/cm^2) is large enough to anneal the contacts, but causes no damage to the tubes. Electrical annealing treatment performed in a cyclic fashion on SWNT bundles of 0.5 μm in length, shows that the overall resistance change of more than 3 orders of magnitude due to contact annealing, bringing the total resistance below 1 kΩ. This method is potentially suitable for short SWNT PNAs. Rapid thermal annealing (RTA) at elevated temperatures (600–800 °C) for few tens of seconds can decrease the contact resistance between a carbon nanotube and metal electrodes (Ti/Au) by several orders of magnitude, becoming stable in the long-term as a result of a better adhesion of the carbon shell to the titanium film [76]. Welding at the nanoscale by a vibrating force at ultrasound frequency has been successful in contacting SWNTs onto metal electrodes, where the formed contact is mechanically robust and it is a stable low-Ohmic with resistances in the range of 8–24 kΩ (for a 1 μm long metallic single wall tubes at room temperature) [77]. For vertical integration, however different approaches have been considered such as pressure contacted soldering method [78], where lithographically patterned carbon nanotubes array grown on silica by chemical vapor deposition (CVD) are transferred of low-temperature (Ag/Sn) coated substrate.Figure 7a,b show the SEM images of carbon nanotubes arrays transferred onto solder pads (positive pattern) using the scheme described on the left as well as the CNTs transferred between the solder pads (negative pattern) in Figure 7c. Infiltration of CNT vias and subsequent polishing in order to expose open MWNT ends is also proposed for such types of architectures [44].

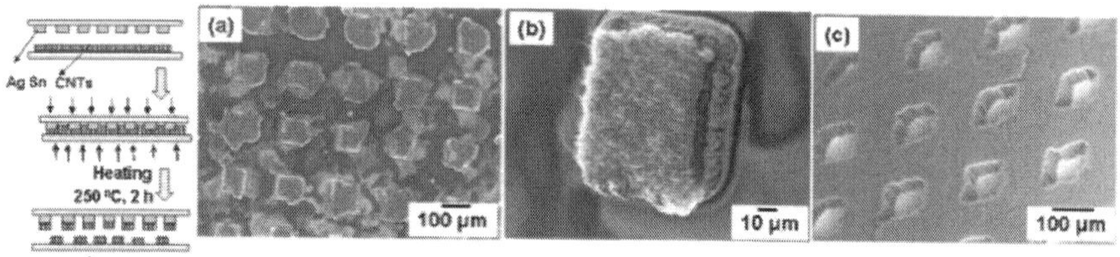

Figure 7. (**a,b**) SEM images of CNTs transferred onto solder pads (positive pattern) using the scheme described on the left; (**c**) CNTs transferred between the solder pads (negative pattern) [78].

Introduction of an interfacial carbon layer has also been proposed in order to reduce the contact resistance as a result of the improvement of the wetting capability and to the formation of chemical bonding to the tube [79]. Also, the contact area is larger due to the presence of the additional carbon layer.

2.2.2. Towards Defect-Free CNTS

Most high-quality CNTs are grown at extremely elevated temperatures (>1000 °C), thus resulting in carbon nanostructures with very low-defect densities. However, in carbon nanotubes grown at lower temperatures, post-growth methods are indeed in need for defects reduction. Ultrafast microwave annealing has been suggested as possible route towards the reduction of defect density in vertically aligned carbon nanotubes, for which only 3-minute-long exposure is necessary to induce a defect reduction, as suggested by Raman spectroscopy and thermogravimetric analysis. Mechanical as well as electric properties are also found to be enhanced upon microwave irradiation [80]. In the presence of a metal catalyst, defects formed during the addition of carbon atoms can be efficiently removed at the tube-catalyst interface, in a self-healing-type of approach. Simulations in fact show that potentially a defect-free CNT wall with 10^8–10^{11} carbon atoms can be achieved, thus leading to a potentially 0.1–100 cm long SWNT [81]. This effect is expected to take place through ring isomerization, leading to the removal of few (5–7) adatoms and monovacancy defects. The timescale of this process is of the order of several picoseconds and depended largely on the catalyst morphology and surface. This healing mechanism also potentially represents a way to control the chirality (n, m) of a growing SWNT during the growth process [82]. Among various catalysts, Fe has demonstrated to be the most efficient in defect healing [81].

Thermal annealing at high temperature has been proposed as an effective method to reduce defects in nanotubes [83,84,85,86]. Annealing at different high temperatures (>1000 °C) of CVD-grown multi-wall carbon nanotubes at relatively low temperatures (650 °C) in porous alumina template lead to a different degree of rearrangement of the imperfect graphitic structure and removal of weakly bounded defects, leading to the reconstruction of the hexagonal carbon lattice along the shell in a more ordered fashion. Transmission electron microscope (TEM) images in Figure 8a demonstrate the increase degree of graphitic order for increased temperatures.

Thermal reduction can also be used to purify and reduce chemical defects in the carbon lattice of single-wall carbon nanotubes, where a combination of first a high-temperature reactions in as-grown nanotubes in a pressurized chamber with a slow annealing process in vacuum environment is used [84]. Application of a rapid large bias across two terminals of a carbon nanotube not only decreases contact resistance, but also can dramatically reduce the number of defects. It has been shown that for highly defective MWNTs [87] (*i.e.*, grown in alumina template), a rapid cycling of the applied voltage across the electrodes of a device generates sufficient Joule heating to anneal defects in the CNT. This can drastically increase the mean free path (by 1–2 orders of magnitude) and reduce the resistance of tubes by orders of magnitude, as shown inFigure 8b [88]. Further, the investigation of the low-temperature charge transport properties allows estimating the defect density in those systems. Certain key transport parameters, such as number of quantum channels, disorder density (or equivalently, the mean free path), and *e-e* interactions play a strong role in determining the electronic properties of the interconnect itself. Using a combination of low-T transport measurement and application of field-theoretical models such as Lüttinger-liquid theory and Al'tshuler-Aronov [87] corrections can be used to estimate the number of quantum channels and mean free path in carbon nanotubes.

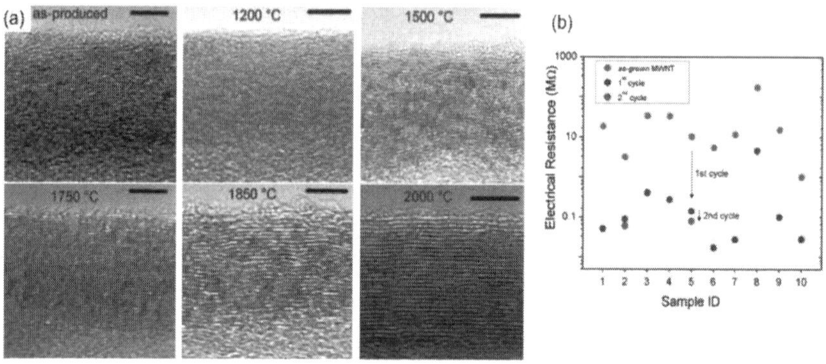

Figure 8. (**a**) Transmission electron micrographs of the walls of as-produced nanotubes grown in alumina template and at different annealing temperatures (as labeled). The initial, disordered carbon structure is converted into a more and more perfect graphitic structure with an increasing annealing temperature. Scale bar is 5 nm; (**b**) Variation of the electrical resistance in individual alumina template grown carbon nanotubes as function of the number of electrical cycles. [85]

2.2.3. Conductance Enhancement via Metal Decoration and Doping

Theoretical and experimental works have shown that modification of the structure of carbon nanotube can drastically affects the electronic properties. In particular, Density Functional Theory (DFT) calculations [89] show that decoration of SWNTs with platinum nanoclusters adds a number of electronic levels and enhances their density of states (DoS) near the Fermi level for both metallic and semiconducting carbon nanotubes, as shown in Figure 9a.

Figure 9b shows the effect of Pt nanocluster decoration on (top) the band-gap and (bottom) the conductance of two semiconducting ((8,0) and (9,0) and a metallic (10,0) single wall nanotube [28]. In case of semiconducting SWNTs, the addition of more than three atoms per cluster sizably brings down its band gap to values very close to zero, and remains close to zero for further decoration. In case of metallic SWNTs, the gap remains closed for values of n larger than 3. Further, calculation of conductance shows that in all three cases, platinum decoration causes the SWNT to approach or remain at the quantum limit of G~4 e^2/h, independently of the tube chirality. Calculations based on more rigorous GW approximation confirm metallic nature of Pt-decorated carbon nanotube [90]. Hence, decoration of carbon nanotubes by metal (Pt) clusters represents a general approach to convert semiconducting nanotubes into metallic, with a consequent effect of improving the overall conductivity of the SWNT PNAs. This process does not require neither selective growth nor post-growth selection of metallic SWNTs for interconnects applications. Encouraging results by electrochemical deposition of minute amounts of platinum on aligned SWNT PNAs have shown that those structures present negligible changes in surface morphology upon metal decoration and a dramatic decrease in resistivity of the order of approximately 75%. Experimentally, it has been demonstrated that decoration of platinum nanoclusters in MWNT has a significant enhancement of conductance. The process of metal decoration is achieved through a high-bias application procedure, which increases the number of quantum channels and decreases the defect density, and their combined effect can be extremely large (several orders of magnitude).

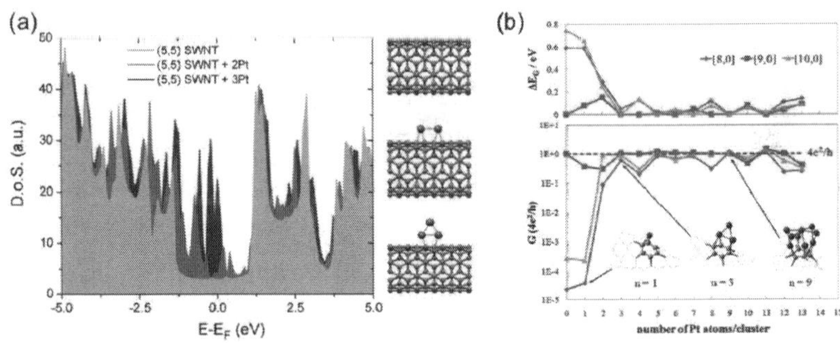

Figure 9. (**a**) Effect of platinum atom decoration on the (top) band-gap and (bottom) conductance of a metallic (9,0) and two semiconducting (8,0) and (10,0) SWNTs. For number of atoms > 3, all semiconducting SWNTs become metallic and their conductance approach the quantum limit of 4 e^2/h. Metallic tube remains metallic; (**b**) Cyclic high-bias treatment (HBT) treatment produced a Pt-nanoclusters decoration of the outer surface of the tube along with a conductance enhancement. (Inset) Platinum-decorated multi wall carbon nanotubes upon HBT [28].

Chemical doping has also been proposed to enhance the conductivity of carbon nanotubes, both as individual objects as well as bundles [91,92,93]. The density of states (DoS) in doped carbon nanotubes reveals the possibility of increasing the conductivity of MWNTs; in fact, chemical doping of the outer shell has the overall effect of shift upwards the Fermi energy, leading to increased conductivity. Being the DoS independent from the tube diameter, elevated conductance values can be expected in case of metallic MWNTs rather than in metallic SWNTs as a results of both larger surface area available for doping and for the fact that any additional conduction channel is close to the Fermi energy in MWNTs. Doping with aryldiazonium and oxonium salts have been shown to be capable of reducing the resistance of MWNTs by almost 20% [94]. Lee *et al.* have shown that bulk samples of SWNTs grown by vapor phase reaction in the presence of bromine and potassium (each one as an example of *e* and *h*-type doping) can lead to an enhancement of the room temperature conductivity of a factor of about 30, which can outperform copper equivalent interconnects [95]. Doping multi wall carbon nanotubes with nitrogen (at low levels) and then performing thermal annealing at temperature of approximately 1000 °C, an elevated decrease in the electrical resistance of individual tubes is observed (about two orders of magnitude), where nitrogen is observed to be stable up to 1500 °C [96]. In case of SWNTs mats, increasing the dopant exposure (Cs) produces first a decrease and then an increase of their electrical resistance, allowing tunability of the dopant level to maximize conductivity of nanotubes structures [97]. In SWNTs mats, one or two order of magnitude increase in conductance can be achieved [95,97,98]. Conduction properties of potassium-doped individual ropes of SWNTs can also be modulated with a gate bias, further showing the intrinsic nature of *h*-doping in pristine tubes [99].

Zhao and coworkers have shown the possibility of fabricating iodine-doped, double-walled nanotube cables having electrical resistivity reaching ~10^{-7} Ω.m. The low-density and their specific density values outperform their copper counterpart and become comparable (one order of magnitude difference) with the highest specific conductivity metal such as sodium. In this case, elevated current handling capacity is also found (10^4–10^5 A/cm^2), along with no degradation of their electrical properties and freedom in terms of size and dimensions growth. Interestingly, a practical application in its preliminary form has been presented by using those doped-nanotubes cables in partly replacing metal connections in a commercially available light bulb circuit [100].

2.2.4. Breakdown in Carbon Nanotubes

One important aspect concerning CNT and their application as interconnects to be considered for realistic exploitation in ICs is the maximum current (density) the tube (or the array/bundle of tubes) can sustain before it breaks or partially loose their electrical properties. Carbon nanotubes both as individual object as well as in arrays/bundles can be tested for breakdown. It is important here to distinguish between single wall and multi wall carbon nanotubes, since the breaking of one single wall in SWNT will lead to the complete breakdown of the tubes, thus failure of the corresponding IC while in the case of MWNTs, there could be multiple and subsequent breakdowns accompanying the breaking of individual shells, leading to variation of the entire conduction properties prior to complete breakdown.

In most cases, breakdown in carbon nanotubes is induced by producing enough Joule heating to completely destroy the continuous carbon shell(s) or, if there are contacts, the contacts themselves. Joule heating can be produced thermally by direct heating and electrically by forcing a current along the tube(s) [101].

In case of single-wall carbon nanotube, electrical breakdown leads to an abrupt drop to zero of the current flowing within the tube, while in the case of multi wall carbon nanotubes and bundles exhibit higher current and multiple breakdown steps. Collins and coworkers have experimentally demonstrated that in arc-grown nanotubes, monitoring the current passing through the tube shows a staircase-shape where each step is associated with the progressive breakdown of adjacent and individual shells (12 μA/ shell) [102]. Similar experimental and theoretical results have been obtained on similar tubes grown by different techniques [18,103,104]. In some cases, the progressive removal of shell in carbon nanotubes, up to complete failure of the tubes, has been shown through electron microscopy (scanning and transmission) [105,106]. As mentioned, breakdown can also be induced thermally by simply heating the tubes. In air conditioning, due to presence of oxygen and self-heating, breakdown temperature is very well known of being approximately 600 °C. Hata and coworkers have carried out experiments both in vacuum (10^{-5} torr) and air and they observed that, in vacuum, carbon nanotubes breakdown takes place at much higher temperatures (approximately 1800 °C) [107]. This implies that oxidation cannot be the only cause leading to breakdown; different works have proposed an important role played by the presence of defects [108,109] and by the underlying substrate and its physical properties (*i.e.*, SiO_2 begin to melt at 1700 °C). Numerous studies have shown that the breakdown in carbon nanotubes is favored and initiated by the presence of oxygen (in air) which tends to etch the outermost shell [102], to the presence of defects along the wall where thus electric field (and consequently heat are intensified) [110,111], as well as to the poor thermal contact between CNTs and metal electrodes which prevents efficient heat removal [112]. It has been observed that metallic carbon nanotubes present the failure point at middle point location [113,114] (most likely corresponding to the hottest point) while semiconducting nanotubes breaks closer to one of the electrodes [111,115] (closer to drain), most likely corresponding to the area of the largest electric field and lower carrier density. Both these results indicate the diffusive nature of the heat dissipation and charge transport within the tube [116,117]. It has been demonstrated that a carbon nanotube grown by CVD with diameter of 18 nm can handle in vacuum a maximum current density up to 8×10^7 A/cm^2 with associated power of 0.6 mW (as compared to 0.16 mW and 2×10^7 A/cm^2 in air) [118]. Decoration with electrochemically deposited Ni particles allows protection of the CNT interconnect against oxidation and improvement of the heat release through the surrounding environment (about 1.5 mW before breaking down under vacuum, with a corresponding maximum current density as high as 1.2×10^8 A/cm^2). The Joule heating melts the nickel particles and promotes the formation of titanium carbon nitride with the overall results of improving the electrical contact between the CNT and the electrodes.

The present Review has so far described various research efforts in the field of fabrication, characterization and development of ultra-high density, super-aligned nanotubes PNA-based high performance horizontal and vertical interconnect assemblies which can be possibly integrated into existing CMOS platforms. Figure 10 shows a typical cross section of an IC circuit (see Figure 1) and potential areas where various CNT-based architectures could begin to replace copper interconnect in the coming years. While it is important to fabricate high-performance CNT PNAs, one has to keep in mind the compatibility and integration considerations in terms of resistivity, failure current density, RC delay and operation bandwidth at appropriate individual scaling levels.

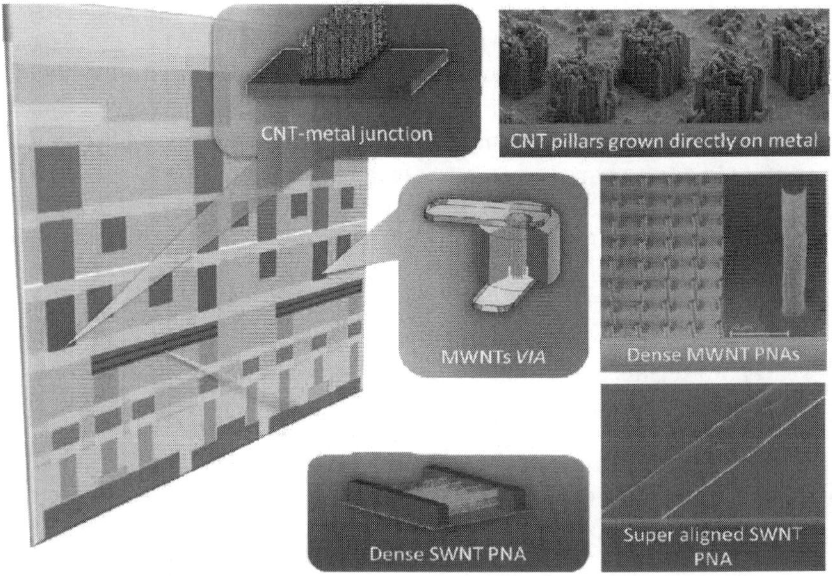

Figure 10. Schematic of a future Integrated Circuit (IC) chip with various interconnects potentially conceivable with CNT-based architectures, as described in the text which can hierarchically enable partial or full integrations in future IC chip [3,28,44,74].

3. GRAPHENE AND GRAPHENE NANORIBBONS

3.1. Graphene as Two-Dimensional Crystal

Upon its experimental evidence in 2004, graphene has soon entered in the quest for the search of a new material for replacement for CMOS-compatible structures [3]. Graphene is a monolayer of *sp2*-bonded carbon atoms packed in a hexagonal lattice, arrangement that makes it rigorously a two-dimensional material [119]. Graphene has gathered the interest of the scientific and technological communities and, the investigation as well as implementation has been exponentially increasing due to the existing technology platform and knowledge, mostly borrowed from the decades-long carbon nanotube community. We refer the Reader to more specific literature for an in-depth and detailed description of the remarkable and interesting properties of this material [120,121]. In the present context, we will focus on some of the properties and characterization that are valuable and of great importance for interconnect applications.

Graphene is a zero bandgap semiconductor, in which electrons behaves as massless Dirac fermions and which could in principle offer less power consumption and enhanced performance due to increased carrier velocities. The ambipolar nature of graphene makes it a very versatile material since it can be used both as an *n*- and *p*-channel material. In fact, charge carriers have been observed to "puddle" in graphene layers at zero field so that both *n*- and *p*-carriers can co-exist within the same layer. Although graphene is a rather new material, a timeline for potential real applications in expected rather soon within the current decade (by 2019–2020 according to International Technology Roadmap for Semiconductors—Emerging Research Materials Section in 2009 [122]).

There exist a variety of methods to produce graphene, ranging from CVD-*like* methods or epitaxial growth on a silicon wafer [123], to mechanical exfoliation [124], chemical oxidation [125] or solvent exfoliation from highly oriented pyrolytic graphite (HOPG) [126,127], direct CVD epitaxy on single crystal metal substrates [128] and sublimation of silicon from SiC [129]. Recently proposed approaches such as growth on polymers (polyethylene terephtalhate, PET or polydimethylsiloxane, PDMS) further provide a new class of flexible and transparent substrates (mechanical bending and twist up to 11% without any loss in conductivity). So far, mechanical exfoliation of graphene has produced high quality films on silicon [119]. Nevertheless, this method is not currently capable of selective placement of graphene at a specific location, which makes it an inadequate route to device fabrication of integrated circuit technologies. Thickness control of several monolayers on a millimeter scale and carrier mobility of 25,000 cm^2/Vs have been demonstrated for grapheme grown on SiC (as a result of Si sublimation), where annealing temperature of at least 1200 °C in H_2 ambient condition are required [130]. Deposition of a thin layer of SiC [131] on a Si wafer followed by silicon evaporation has led to few-atomic layers thin graphene. A common drawback for the above-mentioned techniques is the high temperatures required during processing which might introduce additional defects in 300 mm and 450 mm wafers-line production or might not be compatible with some of the subsequent processes to be performed. Graphene can be grown on single crystal substrates (Ni [132], Ir [133], Pt [134]); however, this process has so far been revealed to be quite expensive. Continuous graphene films have been successfully grown by ambient pressure CVD on polycrystalline Ni which can then be subsequently transferred (by Ni wet-etching) on arbitrary substrates [128], producing films with sheet resistances as small as 280 Ω/□ and mobilities as high as 3700 cm²/Vs. Large-scale graphene growth can be achieved on copper [135] with room temperature electron mobilities of more than 4000 cm^2/Vs. Post-growth transfer of those large-scale graphene domains represents a cheaper (compared to CVD) and effective route for real implementation into industrial standards.

Other techniques such as the solvent exfoliation [127] or the chemical route through production of graphene oxide are currently explored; firstly, graphene can be exfoliated in solvents from HOPG [126] with ultrasound, but solvent choices are very limited because of solvent surface energy thermodynamics. As for graphene oxide, this is a century-long well-established production method consisting first of oxidizing graphene (thermal or chemical intercalation), and then dispersing it in a solvent. However, graphene oxide shows fundamentally different electronic properties than individual graphene layer obtained by mechanical exfoliation or supported growth, thus requiring a chemical reduction step (in hydrazine or hydrogen plasma environment) to retrieve original electronic properties. Solution of chemically reduced grapheme oxide is found to consist primarily of monolayer graphene with mobility between 10 and 1,000 cm^2/Vs, with the conduction dominated by hopping transport through regions of highly ordered graphene surrounded by disordered grapheme [136].

Graphene elevated value of electron mobility at room temperature has rather quickly opened the way to the investigation of a variety of physical phenomenon, ranging from mesoscopic physics [137], Berry's phase [138] and unusual quantum Hall effect [139]. In the case of supported graphene devices, there is a fundamental upper limit of 40,000 cm^2/Vs [140] at room temperature due to trapped charges in the substrate itself [141]. Efficient screening [142] and/or complete removal of the underlying substrate are indeed required in order to enhance the mobility of graphene [143]. Further, it has been shown that in suspended and annealed graphene-based devices, the value of the mobility exceed 200,000 cm^2/Vs, which so far represents the highest value for a semiconductor or a semimetal [144,145]. The mobility is little affected by chemical doping [146] or high electric-field-induced carrier concentrations, suggesting the possibility of achieving rather easily ballistic transport regime on a sub-micrometer scale at 300 K [147].

3.2. Graphene and Few-Graphene Layers Nanoribbons

Although graphene behaves as a semi-metal, it has been shown that it is indeed possible to create an energy gap in graphene through different methods as (i) the inclusion of sp3 hydrocarbon defects in the sp2 lattice [148]; (ii) the distortion of the honeycomb lattice under uniaxial strain [149,150] and (iii) lateral confinement in *ribbon*-like shape [151]. While the first two approaches showing promising results are not seemingly likely to be implemented for interconnect applications, graphene nanoribbons could indeed represent valid and successful candidates [152]. It has been theoretically shown that the electronic properties of graphene nanoribbons can be tuned from perfectly metallic (zero bandgap) [153], in the case of zig-zag edge ribbons (ZGNR), to semiconducting for armchair ribbons (AGNR), where the engineered energy gap varies with the ribbon width and length. The origin of this energy gap in the band-structure of nanoribbons is attributed to edge roughness [151]. In addition, first principle calculations suggest that the carbon-carbon interatomic distance in the vicinity of the edges armchair GNRs is slightly smaller (~3.5% [154]) than the corresponding equivalent distance in two-dimensional graphene (1.42 Å), which can also induce the opening of an energy gap. This can also be favored if the spin component of zigzag GNRs is also considered. It has been recently shown that multilayer armchair GNRs are characterized by different bandgaps, depending on their width as a result of the interlayer interaction between adjacent layer which determines the magnetic polarization and band structures [155]. Various methods have been proposed to fabricate graphene nanoribbons. In particular, *e*-beam lithography allows producing nanoribbons with tailored lateral dimensions starting form graphene sheet, as shown in Figure 11 where the energy gap of NR varies as function of its width. However, these approaches are limited to a ribbon minimum width of approximately 20 nm, along with great difficulties in achieving smooth edges (roughness edges—larger than 5 nm—will affect the electronic properties of the ribbon itself) [151,156]. On the other hand, chemical routes leading to graphene oxide sheets and nanoplatelets can be followed for specific applications; nevertheless, reduction processes are needed to recover conductive properties of graphene [157,158], step that might not be compatible with interconnect technology process.

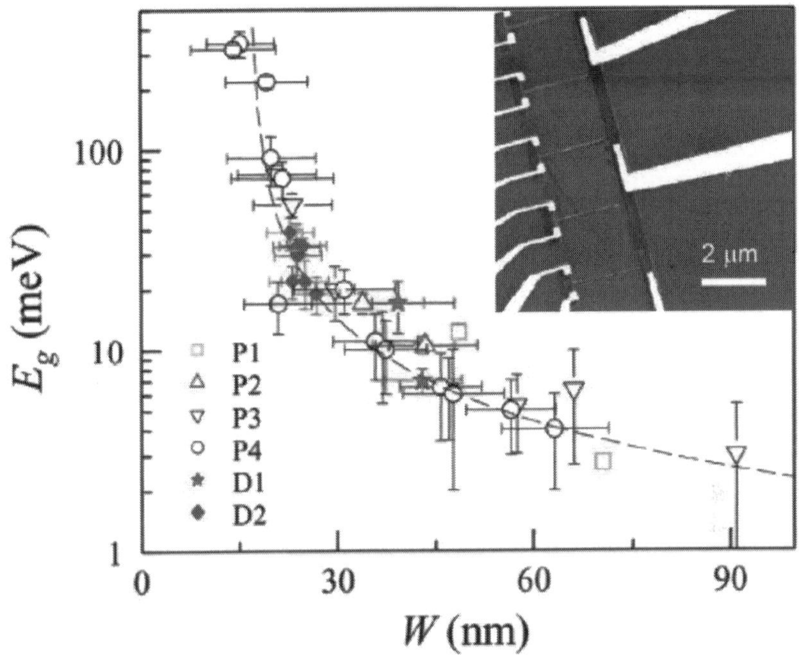

Figure 11. Energy gap dependence from ribbon width in lithographically fabricated graphene nanoribbons. Dashed lines in the inset show the value of as predicted by the empirical scaling of versus W. (Inset) SEM image of a representative graphene nanoribbon-based device [151].

Bottom-up approaches based on the molecular building-up of one-dimensional chains of polyaromatic carbon precursor have been developed [159]; even if this route surely represent a potentially extremely precise method to fabricate graphene nanotibbons with an accurate control over the edge roughness, it is indeed clear that its potential application to large-scale production are not yet ready.

Currently, solution-based methodologies are surely the most promising routes towards the production of graphene nanoribbons, where a very stable compromise between quantity and quality of the ribbons can be reached. Dai and coworkers [160] have shown a simple method to produce GNRs, starting from expandable graphite [161], as shown in Figure 12. Graphite is exfoliated by rapid heating at elevated temperature (1000 °C) in forming gas (Ar/H), dispersed in 1,2-dichloroethane (DCE) solution of poly(*m*-phenylenevinylene-co-2,5-dioctoxy-*p*-phenylenevinylene) and then sonicated. Once a homogeneous solution is obtained, centrifugation is used to separate large and heavy graphite pieces from light and narrow graphene small platelets. Ribbons with different widths down to sub-10 nm region can be thus produced. The same group has lately developed an unconventional and alternative approach based on unzipping mildly oxidized multi-walled carbon nanotubes through mechanical sonication. Quality of these ribbons in terms of disorder and edge roughness is very high, with conductance and mobility of 5 e^2/h and 1500 cm^2/Vs for 10–20 nm long ribbons [162].

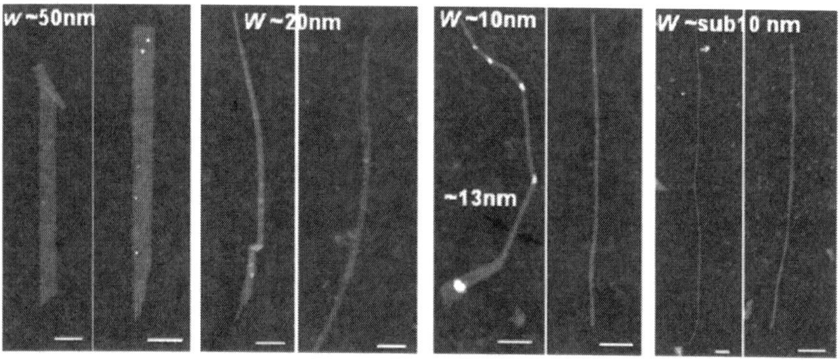

Figure 12. Atomic force microscopy images of selected chemically-derived Graphene Nanoribbons (GNRs) with widths of 50, 30, 20, 10 nm and sub-10-nm, respectively. Scale bar is 100 nm. [160]

Despite the fact that fully metallic GNRs might be hard to produce, this does not rule out the potential of GNRs as interconnects. In fact, semiconducting GNRs have a bandgap smaller than 0.1 eV, value that is equivalent to a copper interconnects with a diameter of 11 nm. If the Fermi energy is larger than half bandgap (few k_BT are sufficient), the first conduction band becomes adequately populated. Hence, for such GNRs, there would be a negligible difference between the ideal conductance of metallic and semiconducting GNRs [163]. Zig-zag and armchair nanoribbons (Fermi Energy = 0.2 eV, interlayer spacing = 0.35 nm) have significantly smaller resistivity than Cu equivalent and densely packed SWNT bundles. For the case of stack of non-interacting GNRs with smooth edges and Fermi energies above 0.2 eV has been predicted to outperform those of copper wires, especially at small cross-sectional dimensions and long lengths [162]. When compared to Cu/low-k interconnects, graphene nanoribbons hold promising potentials for the following reasons: (i) higher conductivity, with electron mean free path as large as a few hundred nanometer [130]; (ii) resistance to electro-migration due to the strong inherent carbon-carbon bonds which accounts for an extraordinary mechanical strength and a very large current conduction capacity of ~10^9A/cm^2 (as compared to ~10^6 A/cm^2 in Cu) and (iii) larger in-plane thermal conductivity of approximately 5,300 W/Km, very much comparable to SWNTs bundles [164]. Nevertheless, there exist important current technological limitations and issues that prevent a much faster implementation for graphene nanoribbons in current interconnect applications.

Wafer-scale synthesis of high-quality graphene sheets on arbitrary substrates remains a major challenge. Successful small- and large-scale growths have been obtained on metals and on SiC substrates; however, the non-dielectric nature of the substrate (SiC) limits its applicability as interconnect. Much needed uniformity and grain size compatible with large-scale integration are indeed still in need. Then, to realistically implement the fabrication of GNRs, the crucial effect that edges play on the overall electronic properties of the GNR has to be taken into consideration. In fact, while the mean free path of non-patterned 2D graphene can be as large as several hundred nanometers, in the case of interaction between conducting electrons and edges, the effective mean free path becomes much smaller and hence, edge scattering becomes an a key issue in the electronic transport at the interconnect length scale. The mean free path associated with edge scattering depends on the edge quality (roughness), GNR width, and the ratio of transverse to longitudinal electron speed. Chemically derived GNRs have been shown to have smoother edges when compared to lithographically patterned ones, with a backscattering probability soon increasing from of 0.2, in the first case to 1 [165]. The product of GNR width and the ratio of longitudinal to transverse velocity determine the average length electrons move before interacting with edges [80]. In addition, doping could be used to shift the Fermi energy of graphene layers from the neutrality as a result of the charge accumulated at the graphene/substrate interface. However, this effect diminishes exponentially for upper layers because of the screening effect [166]. To use all the layers within a multilayer ribbon, the Fermi energy of upper layers must be shifted either through edge functionalization or doping, without any effect on the mean free path of electrons, otherwise, the conductance would become considerably smaller because of their bandgap. This conductance degradation is more severe for narrower GNRs with large bandgap.

In the case of GNRs, contact resistance still represent one of the key aspect/challenge in the real implementation of those structures into real interconnects. In fact, in a FET configuration, source-drain contacts have to be characterized by a low resistance contact, while maintaining unaltered the properties of the ribbon itself (*n*- or *p*-channel nature of the device). As in the case of charge transfer to generate doping, modeling has proposed the use of weak bonding contact metals to generate *n*-type contacts at work functions greater than 5.4 eV [167]. More recently, graphene nanoribbon edge states have been doped *n*-type through high temperature electrochemical ammonia treatment [168]. Experiments have demonstrated that *n*-type graphene can be produced through deposition of NH_3 and CH_4 [169], while *p*-type graphene can be produced through deposition of H_2O and NO_2 [170]. Experiments on graphene ribbons indicate that edges can convert to *p*-type, while the center of the ribbon remains *n*-type [171]. The challenge with these doping techniques will be to maintain the carrier doping in an integrated structure with interconnects. Ohmic contact formation may be easier than in small diameter carbon nanotubes, but a lot still has to be explored in this regard.

Once those numerous technological issues are addressed and overcome, GNRs could potentially be integrated for on-chip interconnect applications in the following two main categories:

(i)*Graphite-like NRs* in place for lower interconnect resistance, especially for short interconnects. In the case of multiple layers, is indeed necessary that the layers in within the stack are yelectronically decoupled [81,120,172], condition which preserves graphene properties [173]. Further, in this case, all layers need to be electrically contacted.

(ii)*Few-layer GNR interconnects* suitable in order to reduce the interconnect capacitance (~ as much as 50%), delay and power dissipation of local interconnects. This arrangement is particularly interesting for short local interconnects in which the delay is dominated by capacitive loading and not resistivity.

4. CONCLUSIONS AND FUTURE CHALLENGES

Carbon-based nanostructures, and in particular carbon nanotubes and graphene nanoribbons, due to their numerous potential and outstanding properties, have rather quickly attracted the interest within different scientific communities in the quest towards replacement of presently used interconnects.

The present Review has provided the reader with a broad overview of such types of materials and architectures which might be suitable to replace current copper/low-k interconnects at various locations onto an integrated circuit. In particular, carbon nanotubes and graphene nanoribbons both hold great potentials at one or more interconnect levels. Although great advances have so far been achieved since the discovery of CNTs in the early 90s and more recently with the experimental evidence of graphene, many challenges and issues are still open regarding their complete exploitation in working devices. Carbon nanotubes and graphene nanoribbons have shown promising results in terms of RC delay, thermal management and conductance, both on individual objects as well as on assemblies in specific architectures; however, much remains to be advanced in order to achieve a full implementation of these architectures into commercially available IC in the next few years.

In fact, while moving towards the 16-nm technology node, crucial bottlenecks are now going to be faced. When downsizing elements below 100 nm size, interconnects face a number of important challenges. As the lateral dimension of interconnects approaches the mean free path of copper the impact of grain boundary scattering, surface scattering, and the presence of a high-resistivity material as a diffusive barrier layer causes a rapid increase in the overall resistivity. Thus, these phenomena cannot be neglected any further. A full integration of CNTs and GNRs in a real MPU requires the precise and strict control of material properties as well as processes to place those elements at desired locations. At the nanoscale, where at times only few hundreds/thousands of atoms are involved, it becomes of crucial importance to understand the nature of interfaces and any possible dynamics taking place at an interface. In this scenario in fact, metal contacts and electrodes on nanoscale objects behave differently when compared to micro- and macroscopic devices.

Precise control over the nature of defects and their density and location within carbon nanotubes and graphene nanoribbons is also significant and of key importance, since the electronic properties are very sensitive to the presence of any defect and their specificity. For example, controlling the edge roughness in the case of graphene ribbons is still an open and challenging issue that, if solved, might open the way to a massive use and exploitation of those materials within realistic architectures. Chirality engineering in carbon nanotubes and controlled growth of high-quality graphene at wafer-scale will surely be further needed towards the full implementation of these materials and their architectures in commercially available devices.

It is also important to note that in the last two decades (and more recently for GNRs), great advances have been achieved in placing CNTs/GNRs at precise location or within predefined arrays (with techniques such as dielectrophoresis or fluidic-based approaches); however, it is mandatory that those methods have to be compatible with any current MPU fabrication process. Further, any proposed materials, whether ultralong carbon nanotubes for long distance interconnects or graphene nanoribbons for short local interconnects, have to retain their properties throughout the entire MPU entire fabrication process.

Last, but not least, the economic impact of the introduction of these new materials in the large scale production of integrated circuits has to be considered. With confidence, the best materials to replace copper/low-*k* interconnects would be the ones that will surely bring innovation and improvement of the performances of both materials and devices. However, to reach a full implementation, at the same time, those materials and technologies will have to have less impact on the infrastructures currently related to the interconnects market.

ACKNOWLEDGEMENTS

CS would like to thank Italian MIUR funded project FIRB RBAP115AYN *Oxides at the nanoscale: multifunctionality and applications* for financial support. ST would like to acknowledge funding support from NSF-ECCS 0925682. SK would like to acknowledge financial support provided by NSF/ECCS-1102481.

CONFLICTS OF INTEREST

The authors declare no conflicts of interest.

REFERENCES

1. International Technology Roadmap for Semiconductors. Interconnect Chapter; SEMATECH: Austin, TX, USA, 2009, 2009 ed. Available online: http://www.itrs.net/Links/2009ITRS/2009Chapters_2009Tables/2009_Interconnect.pdf.

2. Wu, W.; Brongersma, S.H.; van Hove, M.; Maex, K. Influence of surface and grain-boundary scattering on the resistivity of copper in reduced dimensions. Appl. Phys. Lett. 2004, 8, 2838–1.

3. International Technology Roadmap for Semiconductors. Interconnect Chapter; SEMATECH: Austin, TX, USA, 2007, 2007 ed. Available online: http://www.itrs.net/Links/2007ITRS/2007_Chapters/2007_Interconnect.pdf.

4. International Technology Roadmap for Semiconductors, 1999 ed.; In Interconnect ChapterSEMATECH: Austin, TX, USA, 1999.

5. International Technology Roadmap for Semiconductors. Interconnect Chapter; SEMATECH: Austin, TX, USA, 2011, 2011 ed. Available online: http://www.itrs.net/Links/2011ITRS/2011Chapters/2011Interconnect.pdf.

6. Miller, A.B. Optical interconnects to silicon. IEEE J. Sel. Top. Quantum Electron. 2000, 6, 1312–1317.

7. Lu, D. Recent advances on chip-to-chip optical interconnects. Proc. SPIE 2009.

8. Fischbach, R.; Lienig, J.; Meister, T. From 3D Circuit Technologies and Data Structures to Interconnect Prediction. In Proceeding of the 11th International Workshop on System Level Interconnect Prediction, San Francisco, CA, USA, 26–27 July 2009; pp. 77–84.

9. Wei, B.Q.; Vajtai, R.; Ajayan, P.M. Reliability and current carrying capacity of carbon nanotubes. Appl. Phys. Lett. 2001, 79, 1172–1174.

10. Charlier, J.-C.; Blase, X.; Roche, S. Electronic and transport properties of nanotubes. Rev. Mod. Phys.2007, 79, 677–732.

11. Javey, A.; Guo, J.; Wang, Q.; Lundstrom, M.; Dai, H. Ballistic carbon nanotube field-effect transistors.Nature 2003, 424, 654–657.

12. Berber, S.; Kwon, Y.-K.; Tománek, D. Unusually high thermal conductivity of carbon nanotubes. Phys. Rev. Lett. 2000, 84, 4613–4616.

13. Choi, T.-Y.; Poulikakos, D.; Tharian, J.; Sennhauser, U. Measurement of the thermal conductivity of individual carbon nanotubes by the four-point three-method. Nano Lett. 2006, 6, 1589–1593.

14. Close, G.F.; Yasuda, S.; Paul, B.; Fujita, S.; Wong, P.H.S. A 1 GHz integrated circuit with carbon nanotube interconnects and silicon transistors. Nano Lett. 2008, 8, 706–709.

15. Chaste, J.; Lechner, L.; Morfin, P.; Fève, G.; Kontos, T.; Berroir, J.-M.; Glattli, D.C.; Happy, H.; Hakonen, P.; Plaçais, B. Single carbon nanotube transistor at GHz frequency. Nano Lett. 2008, 8, 525–528.

16. Xu, Y.; Srivastava, A. A model for carbon nanotube interconnects. Int. J. Circuit Theory Appl. 2009, 38, 559–575.

17. Srivastava, A.; Xu, Y.; Sharma, A.K. Carbon nanotubes for next generation very large scale integration interconnects. J. Nanophotonics 2010, 4, 041690:1–041690:27.

18. Frank, S.; Poncharal, P.; Wang, Z.L.; de Heer, W.A. Carbon nanotube quantum. Resist. Sci. 1998, 280, 1744–1746.

19. Van der Straten, O.; Zhy, Y.; Eisenbraun, E.; Kaloyeros, A. Thermal and Electrical Barrier Performance Testing of Ultrathin Atomic Layer Deposition Tantalum-Based Materials for Nanoscale Copper Metallization. In Proceedings of the IEEE International Interconnect Technology Conference, Burlingame, CA, USA, 5 June 2002; pp. 188–190.

20. Watanabe, T.; Nasu, H.; Minamihaba, G.; Kurashima, N.; Gawase, A.; Shimada, M.; Yoshimizu, Y.; Uozumi, Y.; Shibata, H. Self-Formed Barrier Technology using CuMn Alloy Seed for Cu Dual-Damascene Interconnect with Porous-SiOC/ Porous-Par Hybrid Dielectric. In Proceedings of the IEEE International Interconnect Technology Conference, Burlingame, CA, USA, 4–6 June 2007; pp. 7–9.

21. Zhou, Y.; Sreekala, S.; Ajayan, P.M.; Nayak, S.K. Resistance of copper nanowires and comparison with carbon nanotube bundles for interconnect applications using first principles

calculations. J. Phys. Condens. Matter 2008, 20, 095209:1–095209:9.

22. Maffucci, A.; Miano, G.; Villone, F. Performance comparison between metallic carbon nanotube and copper nano-interconnects. IEEE Trans. Adv. Packag. 2008, 31, 692–699.

23. Cho, H.; Koo, K.-H.; Kapur, P.; Saraswat, K.C. Performance comparisons between Cu/low-κ, carbon-nanotube, and optics for future on-chip interconnects. IEEE Electron Device Lett. 2008, 29, 122–124.

24. Naeemi, A.; Meindl, J.D. Design and performance modeling for single-walled carbon nanotubes as local, semiglobal, and global interconnects in gigascale integrated systems. IEEE Trans. Electron Devices 2007, 54, 26–37.

25. Naeemi, A.; Meindl, J.D. Compact physical models for multiwall carbon-nanotube interconnects. IEEE Electron Device Lett. 2006, 27, 338–340.

26. Naeemi, A.; Meindl, J.D. Monolayer metallic nanotube interconnects: Promising candidates for short local interconnects. IEEE Electron Device Lett. 2005, 26, 544–546.

27. Naeemi, A.; Meindl, J.D. Performance comparison between carbon nanotube and copper interconnects for gigascale integration (GSI). IEEE Electron Device Lett. 2005, 26, 84–86.

28. Kim, Y.L.; Li, B.; An, X.; Hahm, M.G.; Chen, L.; Washington, M.; Ajayan, P.M.; Nayak, S.K.; Busnaina, A.; Kar, S.; Jung, Y.J. Highly aligned scalable platinum-decorated single-wall carbon nanotube arrays for nanoscale electrical interconnects. ACS Nano 2009, 3, 2818–2826.

29. Ohashi, T.; Ochiai, T.; Tokune, T.; Kawarada, H. Increasing the length of a single-wall carbon nanotube forest by adding titanium to a catalytic substrate. Carbon 2013, 57, 79–87.

30. Peng, H.B.; Ristroph, T.G.; Schurmann, G.M.; King, G.M.; Yoon, J.; Narayanamurti, V.; Golovchencko, J.A. Patterned growth of single-walled carbon nanotube arrays from a vapor-deposited Fe catalyst.Appl. Phys. Lett. 2003, 83, 4238–4240.

31. Jin, Z.; Chu, H.; Wang, J.; Hong, J.; Tan, W.; Li, Y. Ultralow feeding gas flow guiding growth of large-scale horizontally aligned

single-walled carbon nanotube arrays. Nano Lett. 2007, 7, 2073–2079.

32. Shekhar, S.; Stokes, P.; Khondaker, S.I. Ultrahigh density alignment of carbon nanotube arrays by dielectrophoresis. ACS Nano 2011, 5, 1739–146.

33. Kang, S.J.; Kocabas, C.; Kim, H.-S.; Cao, Q.; Meitl, M.A.; Khang, D.-Y.; Rogers, J.A. Printed multilayer superstructures of aligned single-walled carbon nanotubes for electronic applications. Nano Lett. 2007,7, 3343–3348.

34. Han, S.-P.; Maune, H.T.; Barish, R.D.; Bockrath, M.; Goddard, W.A. DNA-inker-induced surface assembly of ultra dense parallel single walled carbon nanotube arrays. Nano Lett. 2012, 12, 1129–1135.

35. Hong, L.; Wen-Yan, Y.; Banerjee, K.; Jun-Fa, M. Circuit modeling and performance analysis of multi-walled carbon nanotube interconnects. IEEE Trans. Electron Devices 2008, 55, 1328–1337.

36. Punbusayakul, N.; Ci, L.; Talapatra, S.; Surareungchai, M.; Ajayan, P.M. Ultralong aligned multi-walled carbon nanotube for electrochemical sensing. J. Nanosci. Nanotechnol. 2008, 8, 2085–2090.

37. Li, X.; Ci, L.; Kar, S.; Soldano, C.; Kilpatrick, S.J.; Ajayan, P.M. Densified aligned carbon nanotube films via vapor phase infiltration of carbon. Carbon 2007, 45, 847–851.

38. Hong, B.H.; Lee, J.Y.; Beetz, T.; Zhu, Y.; Kim, P.; Kim, K.S. Quasi-continuous growth of ultralong carbon nanotube arrays. J. Am. Chem. Soc. 2005, 127, 15336–15337.

39. Li, X.; Zhang, X.; Ci, L.; Shah, R.; Wolfe, C.; Kar, S.; Talapatra, S.; Ajayan, P.M. Air-assisted growth of ultra-long carbon nanotube bundles. Nanotechnology 2008, 19, 455609.

40. Li, X.; Cao, A.; Jung, Y.J.; Vajtai, R.; Ajayan, P.M. Bottom-up growth of carbon nanotube multilayers: Unprecedented growth. Nano Lett. 2005, 5, 1997–2000.

41. Jackson, J.J.; Puretzky, A.A.; More, K.L.; Rouleau, C.M.; Eres, G.; Geohegan, D.B. Pulsed growth of vertically aligned nanotube arrays with variable density. ACS Nano 2010, 4, 7573–7581.

42. Esconjauregui, S.; Fouquet, M.; Bayer, B.C.; Ducati, C.; Smajda, R.; Hofmann, S.; Robertson, J. Growth of ultrahigh density vertically aligned carbon nanotube forests for interconnects. ACS Nano 2010, 4, 7431–7436.

43. Engel, M.; Small, J.P.; Steiner, M.; Freitag, M.; Green, A.A.; Hersam, M.C.; Avouris, Ph. Thin film nanotube transistors based on self-assembled, aligned, semiconducting carbon nanotube arrays. ACS Nano 2008, 2, 2445–2452.

44. Liu, Z.; Ci, L.; Kar, S.; Ajayan, P.M.; Lu, J.-Q. Fabrication and electrical characterization of densified carbon nanotube micro-pillars for IC interconnection. IEEE Trans. Nanotechnol. 2009, 8, 196–203.

45. Phok, S.; Rajaputra, S.; Sing, V.P. Copper indium diselenide nanowire arrays by electrodeposition in porous alumina templates. Nanotechnology 2007, 18, 475601:1–475601:8.

46. Saito, T.; Ohshima, S.; Okazaki, T.; Ohmori, S.; Yumura, M.; Iijima, S. Selective diameter control of single-walled carbon nanotubes in the gas-phase synthesis. J. Nanosci. Nanotechnol. 2008, 8, 6153–6157.

47. Hiraoka, T.; Bandow, S.; Shinohara, H.; Iijima, S. Control on the diameter of single-walled carbon nanotubes by changing the pressure in floating catalyst CVD. Carbon 2006, 44, 1853–1859.

48. Wang, Y.; Liu, Y.; Li, X.; Cao, L.; Wei, D.; Zhang, H.; Shi, D.; Yu, G.; Kajiura, H.; Li, Y. Direct enrichment of metallic single-walled carbon nanotubes induced by the different molecular composition of monohydroxy alcohol homologues. Small 2007, 3, 1486–1490.

49. Harutyunyan, A.R.; Chen, G.; Paronyan, T.M.; Pigos, E.M.; Kuznetsov, O.A.; Hewaparakrama, K.; Kim, S.M.; Zakharov, D.; Stach, E.A.; Sumanasekera, G.U. Preferential growth of single-walled carbon nanotubes with metallic conductivity. Science 2009, 326, 116–120.

50. He, M.; Chernov, A.I.; Fedotov, P.V.; Obraztsova, E.D.; Rikkinen, E.; Zhu, Z.; Sainio, J.; Jiang, H.; Nasibulin, A.G.; Kauppinen, E.I.; Niemelä, M.; Krause, A.O.I. Selective growth of SWNTs on partially reduced monometallic cobalt catalyst. Chem. Commun. 2011, 47, 129–1221.

51. He, M.; Jiang, H.; Liu, B.; Fedotov, P.V.; Chernov, A.I.; Obraztsova, E.D.; Cavalca, F.; Wagner, J.B.; Hansen, T.W.; Anoshkin, I.V.; et al. Chiral-selective growth of single-walled carbon nanotubes on lattice-mismatched epitaxial cobalt nanoparticles. Sci. Rep. 2013.

52. Voggu, R.; Gosh, S.; Govindaraj, A.; Rao, C.N.R. New strategies for the enrichment of metallic single-walled carbon nanotubes. J. Nanosci. Nanotechnol. 2010, 10, 4102–4108.

53. Fu, Y.; Chen, S.; Bielecki, J.; Matic, A.; Wang, T.; Ye, L.-L.; Liu, J. Selective growth of double walled carbon nnaotubes on gold films. Mater. Lett. 2012, 72, 78–80.

54. Yamada, T.; Namai, T.; Hata, K.; Futaba, D.N.; Mizuno, K.; Fan, J.; Yudasaka, M.; Yumura, M.; Iijima, S. Size-selective growth of double-walled carbon nanotube forests from engineered iron catalysts. Nat. Nanotechnol. 2006, 1, 131–136.

55. Koshio, A.; Yudasaka, M.; Iijima, S. Disappearance of inner tubes and generation of double-wall carbon nanotubes from highly dense multiwall carbon nanotubes by heat treatment. J. Phys. Chem. C2007, 111, 10–12.

56. Lastella, S.; Mallick, G.; Woo, R.; Karna, S.P.; Rider, D.A.; Manners, I.; Jung, Y.J.; Ryu, C.Y.; Ajayan, P.M. Parallel arrays of individually addressable single-walled carbon nanotube field-effect transistors.J. Appl.Phys. 2006, 99, 024302:1–024302:4.

57. Collins, P.G.; Arnold, M.S.; Avouris, P. Engineering carbon nanotubes and nanotube circuits using electrical breakdown. Science 2001, 292, 706–709.

58. Arnold, M.S.; Green, A.A.; Hulvat, J.F.; Stupp, S.I.; Hersam, M.C. Sorting carbon nanotubes by electronic structure using density differentiation. Nat. Nanotechnol. 2006, 1, 60–65.

59. Bonaccorso, F.; Hasan, T.; Tan, P.H.; Sciascia, C.; Privitera, G.; di Marco, G.; Gucciardi, P.G.; Ferrari, A.C. Density gradient ultracentrifugation of nanotubes: Interplay of bundling and surfactants encapsulation. J. Phys. Chem. C 2010, 114, 17267–17285.

60. Ghosh, S.; Bachilo, S.M.; Weisman, R.B. Advanced sorting of single-walled carbon nanotubes by nonlinear density-gradient ultracentrifugation. Nat. Nanotechnol. 2010, 5, 443–450.

61. Krupke, R.; Hennrich, F.; von Lohneysen, H.; Kappes, M.M. Separation of metallic from semiconducting single-walled carbon nanotubes. Science 2003, 301, 344–347.

62. Nieuwoudt, A.; Massoud, Y. Evaluating the impact of resistance in carbon nanotube bundles for VLSI interconnect using diameter-dependent modeling techniques. IEEE Trans. Electron Devices 2006, 53, 2460–2466.

63. Su, Y.; hang, Y.; Wei, H.; Yang, Z.; Kong, E. S.-W.; Zhang, Y. Diameter-control of single-walled carbon nanotubes produced by magnetic field-assisted arc discharge. Carbon 2012, 50, 2556–2562.

64. Ural, A.; Li, Y.M.; Dai, H.J. Electric-field-aligned growth of single-walled carbon nanotubes on surfaces. Appl. Phys. Lett. 2002, 81, 3464:1–3464:3.

65. Huang, S.; Cai, X.; Liu, J. Growth of millimeter-long and horizontally aligned single-walled carbon nanotubes on flat substrates. J. Am. Chem. Soc. 2003, 125, 5636–5637.

66. Ismach, A.; Segev, L.; Wachtel, E.; Joselevich, E. Atomic-step-templated formation of single wall carbon nanotube patterns. Angew. Chem. 2004, 116, 6266–6269.

67. Kocabas, C.; Hur, S.H.; Gaur, A.; Meitl, M.A.; Shib, M.; Rogers, J.A. Guided growth of large-scale, horizontally aligned arrays of single-walled carbon nanotubes and their use in thin-film transistors.Small 2005, 1, 1110–1116.

68. Wei, Y.Y.; Eres, G.; Merkulov, V.I.; Lowndes, D.H. Effect of catalyst film thickness on carbon nanotube growth by selective area chemical vapor deposition. Appl. Phys. Lett. 2001, 78, 1394:1–1394:3.

69. Chai, Y.; Xiao, Z.; Chan, P.C.H. Horizontally aligned carbon nanotube bundles for interconnect application: Diameter-dependent contact resistance and mean free path. Nanotechnology 2010, 21, 235705.

70. Chiodarelli, N.; Fournier, A.; Okuno, H.; Dijon, J. Carbon nanotubes horizontal interconnects with end-bonded contacts, diameters down to 50 nm and lengths up to 20 μm. Carbon 2013, 60, 139–145.

71. Yan, F.; Zhang, C.; Cott, D.; Zhong, G.; Robertson, J. High-density growth of horizontally aligned carbon nanotubes for interconnects. Phys. Status Solidi B 2010, 247, 2669–2672.

72. Zhang, C.; Yan, F.; Allen, C.S.; Bayer, B.C.; Hofmann, S.; Hickey, B.J.; Cott, D.; Zhong, G.; Robertson, J. Growth of vertically-aligned carbon nanotube forests on conductive cobalt disilicide support. J. Appl. Phys. 2010, 108, 024311:1–024311:6.

73. Hiraoka, T.; Yamada, T.; Hata, K.; Futaba, D.N.; Kurachi, H.; Uemura, S.; Yumura, M.; Iijima, S. Synthesis of single- and double-walled carbon nanotube forests on conducting metal foils. J. Am. Chem. Soc. 2006, 128, 13338–13339.

74. Talapatra, S.; Kar, S.; Pal, S.; Vajtai, R.; Ci, L.; Victor, P.; Shaijumon, M.M.; Kaur, S.; Nalamasu, O.; Ajayan, P.M. Direct growth of aligned carbon nanotubes on bulk metals. Nat. Nanotechnol. 2006, 1, 112–116.

75. Pal, S.; Talapatra, S.; Kar, S.; Ci, L.; Vajtai, R.; Borca-Tasciuc, T.; Schadler, L.S.; Ajayan, P.M. Time and temperature dependence of multi-walled carbon nanotube growth on Inconel 600. Nanotechnology2008, 19, 045610.

76. Lee, J.-O.; Park, C.; Kim, J.J.; Kim, J.; Park, J.W.; Yoo, K.H. Formation of low-resistance ohmic contacts between carbon nanotube and metal electrodes by a rapid thermal annealing method. J. Phys. D Appl. Phys. 2000, 33, 1953–1956.

77. Chen, C.; Yan, L.; Kong, E.S.-W.; Zhang, Y. Ultrasonic nanowelding of carbon nanotubes to metal electrodes. Nanotechnology 2006, 17, 2192–2197.

78. Kumar, A.; Pushparaj, V.L.; Kar, S.; Nalamasu, O.; Ajayan, P.M.; Baskaran, R. Contact transfer of aligned carbon nanotube arrays onto conducting substrates. Appl. Phys. Lett. 2006, 89, 163120:1–163120:3.

79. Chai, Y.; Hazegi, A.; Takei, K.; Chen, H.-Y.; Chan, P.C.H.; Javey, A.A.; Wong, H.-S.-S.P. Low-resistance electrical contact to carbon nanotubes with graphitic interfacial layer. IEEE Trans. Electron. Devices2012, 59, 12–19.

80. Lin, W.; Moon, K.-S.; Zhang, S.; Ding, Y.; Shang, J.; Chen, M.; Wong, C.-P. Microwave makes carbon nanotubes less defective. ACS Nano 2010, 4, 1716–1722.

81. Yuan, Q.; Xu, Z.; Yacobson, B.I.; Ding, F. Efficient defect healing in catalytic carbon nanotube growth.Phys. Rev. Lett. 2012, 108, 245505:1–245505:5.

82. Page, A.J.; Ohta, Y.; Okamoto, Y.; Irle, S.; Morokuma, K. Defect healing during single-walled carbon nanotube growth: A density-functional tight-binding molecular dynamics investigation. J. Phys. Chem. C 2009, 113, 20198–20207.

83. Tsai, J.T.H.; Tseng, A.A. Defect reduction of multi-walled carbon nanotubes by rapid vacuum arc annealing. J. Exp.Nanosci. 2009, 4, 87–93.

84. Tran, N.E.; Lambrakos, S.G. Purification and defect elimination of single-walled carbon nanotubes by the thermal reduction technique. Nanotechnology 2005, 16, 639–646.

85. Mattia, D.; Rossi, M.P.; Kim, B.M.; Korneva, G.; Bau, H.H.; Gogotsi, Y. Effect of graphitization on the wettability and electrical conductivity of CVD-carbon nanotubes and films. J. Phys. Chem. B 2006, 110, 9850–9855.

86. Andrews, R.; Jacques, D.; Qian, D.; Dickey, E.C. Purification and structural annealing of multiwalled carbon nanotubes at graphitization temperatures. Carbon 2001, 39, 1681–1687.

87. Kar, S.; Soldano, C.; Chen, L.; Talapatra, S.; Vajtai, R.; Nayak, S.K.; Ajayan, P.M. Lüttinger iquid to Al'tshuler-Aronov transition in disordered, many-channel carbon nanotubes. ACS Nano 2009, 3, 207–212.

88. Soldano, C. Low-Temperature Electronic Transport in One-Dimensional Hybrid Systems: Metal Cluster Embedded Carbon Nanotubes. Ph.D. Thesis, Rensselaer Polytechnic Institute, Troy, NY, USA, 2007.

89. Parr, R.G.; Yang, W. Density-Functional Theory of Atoms and Molecules; Oxford University Press: Oxford, UK, 1989.

90. Hybertsen, M.S.; Louie, S.G. First-principles theory of quasiparticles: Calculation of band gaps in semiconductors and insulators. Phys. Rev. Lett. 1985, 55, 1418–1421.

91. Ryu, Y.; Yin, L.; Yu, C. Dramatic electrical conductivity improvement of carbon nanotube networks by simultaneous de-bundling and hole-doping with chlorosulfonic acid. J. Mater.Chem. 2012, 22, 6959–6964.

92. Zhou, C.; Wang, S.; Zhuang, Q.; Han, Z. Enhanced conductivity in polybenzoxazoles doped with carboxylated multi-walled carbon nanotubes. Carbon 2008, 46, 1232–1240.

93. Bockrath, M.; Hone, J.; Zettl, A.; McEuen, P.L.; Rinzler, A.G.; Smalley, R. Chemical doping of individual semiconducting carbon-nanotube ropes. Phys. Rev. B 2000, 61, R10606–R10608.

94. Neumann, A. Chemical Optimization of Carbon Nanotubes for Electronic Devices. Diploma Thesis, Technical University of Ilmenau, Ilmenau, Germany, 2005.

95. Lee, R.S.; Kim, H.J.; Fischer, J.E.; Thess, A.; Smalley, R.E. Conductivity enhancement in single-walled carbon nanotube bundles doped with K and Br. Nature 1997, 388, 255–257.

96. Fujisawa, K.; Tojo, T.; Muramatsu, H.; Elías, A.L.; Vega-Díaz, S.M.; Tristán-López, F.; Kim, J.H.; Hayashi, T.; Kim, Y.A.; Endo, M.; et al. Enhanced electrical conductivities of N-doped carbon nanotubes by controlled heat treatment. Nanoscale 2011, 3, 4359–4364.

97. Grigorian, L.; Sumanasekera, G.U.; Loper, A.L.; Fang, S.; Allen, J.L.; Eklund, P.C. Transport properties of alkali-metal-doped single-wall carbon nanotubes. Phys. Rev. B 1998, 58, R4195–R4198.

98. Ruzicka, B.; Degiorgi, L.; Gaal, R.; Thien-Nga, L.; Bacsa, R.; Salvetat, J.-P.; Forrò, L. Optical and dc conductivity study of potassium-doped single-walled carbon nanotube films. Phys. Rev. B 2000, 61, R2468–R2471.

99. Lee, R.S.; Kim, H.J.; Fischer, J.E.; Lefebvre, J.; Radosavljević, M.; Hone, J.; Johnson, A.T. Transport properties of a potassium-doped single-wall carbon nanotube rope. Phys. Rev. B 2000, 61, 4526–4529.

100. Zhao, Y.; Wei, J.; Vajtai, R.; Ajayan, P.M.; Barrera, E.V. Iodine doped carbon nanotube cables exceeding specific electrical conductivity of metals. Sci. Rep. 2011, 1, 83.

101. Mohsin, K.M.; Srivastava, A.; Sharma, A.K.; Mayberry, C. A thermal model for carbon nanotube interconnects. Nanomaterials 2013, 3, 229–241.

102. Collins, P.G.; Hersam, M.; Arnold, M.; Martel, R.; Avouris, P. Current saturation and electrical breakdown in multiwalled carbon nanotubes. Phys. Rev. Lett. 2001, 86, 3128–3131.

103. Collins, P.G.; Avouris, P. Multishell conduction in multiwalled carbon nanotubes. Appl. Phys. A 2002,74, 329–332.

104. Dai, H.J.; Wong, E.W.; Lieber, C.M. Probing electrical transport in nanomaterials: Conductivity of individual carbon nanotubes. Science 1996, 272, 523–526.

105. Subramanian, A.; Dong, L.; Frutiger, D.; Nelson, B.J. Shell Engineering of Carbon Nanotube Arrays by Current Driven Breakdown. In IEEE-NANO 2006,In Proceedings of Sixth IEEE Conference on Nanotechnology, Cincinnati, OH, USA, 17–20 July 2006; Volume 2, pp. 901–904.

106. Lim, S.C.; Kim, K.S.; Lee, I.B.; Jeong, S.Y.; Cho, S.; Yoo, J.-E.; Lee, Y.H. Nanomanipulator-assisted fabrication and characterization of carbon nanotubes inside scanning electron microscope. Micron2005, 36, 471–476.

107. Hata, K.; Futaba, D.N.; Mizuno, K.; Namai, T.; Yumura, M.; Iijima, S. Water-assisted highly efficient synthesis of impurity-free single-walled carbon nanotubes. Science 2004, 306, 1362–1364.

108. Huang, N.Y.; She, J.C.; Chen, J.; Deng, S.Z.; Xu, N.S.; Bishop, H.; Huq, S.E.; Wang, L.; Zhong, D.E.; Wang, E.G.; et al. Mechanism responsible for initiating carbon nanotube vacuum breakdown. Phys. Rev. Lett. 2004, 93, 075501:1–075501:4.

109. Yuzvinsky, T.D.; Mickelson, W.; Aloni, S.; Begtrup, G.E.; Kis, A.; Zettl, A. Shrinking a carbon nanotube. Nano Lett. 2006, 6, 2718–2722.

110. Molhave, K.; Gudnason, S.B.; Pedersen, A.T.; Clausen, C.H.; Horsewell, A.; Boggild, P. Transmission electron microscopy study of individual carbon nanotube breakdown caused by Joule heating in air.Nano Lett. 2006, 6, 1663–1668.

111. Berger, C.; Yi, Y.; Gezo, J.; Poncharal, P.; de Heer, W.A. Contacts, non-linear transport effects and failure in multi-walled carbon nanotubes. New J. Phys. 2003, 5, 1–16.

112. Pop, E. Energy dissipation and transport in nanoscale devices. Nano Res. 2010, 3, 147–169.

113. Lee, Y.H.; Lee, J.H.; Chung, S.J.; Lee, S.; Ju, B.K. Carrier carrying capacity of one-step grown suspended carbon nanotube bridge with carbon nanotube contact electrodes: For practical one-dimensional electronics. Appl. Phys. Lett. 2006, 89, 07310:1–07310:3.

114. Yuzvinsky, T.D.; Mickelson, W.; Aloni, S.; Konsek, S.L.; Fennimore, A.M.; Begtrup, G.E., Kis; Regan, B.C.; Zettl, A. Imaging the life story of nanotube devices. Appl. Phys. Lett. 2005, 87, 083103:1–083103:3.

115. Wei, Y.; Liu, P.; Jiang, K.L.; Liu, L.; Fan, S.S. Breaking single-walled carbon nanotube bundles by Joule heating. Appl. Phys. Lett. 2008, 93, 023118:1–023118:3.

116. Pop, E.; Mann, D.A.; Goodson, K.E.; Dai, H.J. Electrical and thermal transport in metallic single-wall carbon nanotubes on insulating substrates. J. Appl. Phys. 2007, 101, 093710:1–093710:10.

117. Pop, E.; Mann, D.; Wang, Q.; Goodson, K.; Dai, H.J. Thermal conductance of an individual single-wall carbon nanotube above room temperature. Nano Lett. 2006, 6, 96–100.

118. Santini, C.A.; Vereecken, P.M.; Volodin, A.; Groeseneken, G.; de Gendt, S.; van Haesendonck, C. A study of Joule heating-induced breakdown of carbon nanotube interconnects. Nanotechnology 2011, 22, 395202:1–395202:9.

119. Geim, A.K.; Novoselov, K.S. The rise of graphene. Nat. Mater. 2007, 6, 183–191.

120. Soldano, C.; Mahmood, A.; Dujardin, E. Production, properties and potential of graphene. Carbon2010, 48, 2127–2150.

121. Zhu, Y.; Murali, S.; Cai, W.; Li, X.; Suk, J.W.; Potts, J.R.; Ruoff, R.S. Graphene and graphene oxide: Synthesis, properties, and applications. Adv. Mater. 2010, 22, 3906–3924.

122. International Technology Roadmap for Semiconductors. Emerging Research Materials Chapter; SEMATECH: Austin, TX, USA, 2009, 2009 ed. Available online: http://www.itrs.net/Links/2009Chapters_2009Tables/2009_ERM.pdf.

123. Kim, K.S.; Zhao, Y.; Jang, H.; Lee, S.Y.; Kim, J.M.; Kim, K.S.; Ahn, J.-H.; Kim, P.; Choi, J.-Y.; Hong, B.H. Large-scale pattern growth of

graphene films for stretchable transparent electrodes. Nature 2009, 457, 706–710.

124. Novoselov, K.S.; Geim, A.K.; Morozov, S.V.; Jiang, D.; Zhang, Y.; Dubonos, S.V.; Grigorieva, I.V.; Firsov, A.A. Electric field effect in atomically thin carbon films. Science 2004, 306, 666–669.

125. Brodie, B.C. On the atomic weight of graphite. Philos. Trans. R. Soc. Lond. 1859, 149, 249–259.

126. Hernandez, Y.; Nicolosi, V.; Lotya, M.; Blighe, F.M.; Sun, Z.; De, S.; McGovern, I.T.; Holland, B.; Byrne, M.; Gun'Ko, Y.K.; et al. High-yield production of graphene by liquid-phase exfoliation of graphite.Nat. Nanotechnol. 2008, 3, 563–568.

127. An, X.; Simmons, T.; Shah, R.; Wolfe, C.; Lewis, K.M.; Washington, M.; Nayak, S.K.; Talapatra, S.; Kar, S. Stable aqueous dispersions of noncovalently functionalized graphene from graphite and their multifunctional high-performance applications. Nano Lett. 2010, 10, 4295–4301.

128. Reina, A.; Jia, X.; Ho, J.; Nezich, D.; Son, H.; Bulovic, V.; Dresselhaus, M.S.; Kong, J. Large area, few-layer graphene films on arbitrary substrates by chemical vapor deposition. Nano Lett. 2009, 9, 30–35.

129. Berger, C.; Song, Z.M.; Li, T.B.; Li, X.B.; Ogbazghi, A.Y.; Feng, R.; Dai, Z.; Marchenkov, A.N.; Conrad, E.H.; First, P.N.; et al. Ultrathin epitaxial graphite: 2D electron gas properties and a route toward graphene-based nanoelectronics. J. Phys. Chem. B 2004, 108, 19912–19916.

130. Berger, C.; Song, Z.; Li, X.; Wu, X.; Brown, N.; Naud, C.; Mayou, D.; Li, T.; Hass, J.; Marchenkov, A.N.;et al. Electronic confinement and coherence in patterned epitaxial graphene. Science 2006, 312, 1191–1196.

131. Suemitsu, M.; Miyamoto, Y.; Handa, H.; Konno, A. Graphene formation on a 3C-SiC(111) thin film grown on Si(110) substrate. e-J. Surf. Sci. Nanotechnol. 2009, 7, 311–313.

132. Gamo, Y.; Nagashima, A.; Wakabayashi, M.; Terai, M.; Oshima, C. Atomic structure of monolayer graphite formed on Ni(111). Surf. Sci. 1997, 374, 61–64.

133. Coraux, J.; N'Diaye, A.T.; Engler, M.; Busse, C.; Wall, D.; Buckanie, N.; Meyer zu Heringdorf, F.J.; van Gastel, R.; Poelsema, B.;

Michely, T. Growth of graphene on Ir(111). New J. Phys. 2009, 11, 023006:1–023006:25.

134. Preobrajenski, A.B.; Ling Ng, M.; Vinogradov, A.S.; Mårtensson, N. Controlling graphene corrugation on lattice-mismatched substrates. Phys. Rev. B 2008, 78, 073401.

135. Li, X.; Cai, W.; An, J.; Kim, S.; Nah, J.; Yang, D.; Piner, R.; Velamakanni, A.; Jung, I.; Tutuc, E.; et al. Large-area synthesis of high-quality and uniform graphene films on copper foils. Science 2009, 324, 1312–1314.

136. Gómez-Navarro, C.; Weitz, R.T.; Bittner, A.M.; Scolari, M.; Mews, A.; Burghard, M.; Kern, K. Electronic transport properties of individual chemically reduced graphene oxide sheets. Nano Lett. 2007, 7, 3499–3503.

137. Novoselov, K.S.; Geim, A.K.; Morozov, S.V.; Jiang, D.; Katsnelson, M.I.; Grigorieva, I.V.; Dubonos, S.V.; Firsov, A.A. Two-dimensional gas of massless Dirac fermions in graphene. Nature 2005, 438, 197–200.

138. Zhang, Y.B.; Tan, Y.W.; Stormer, H.L.; Kim, P. Experimental observation of the quantum Hall effect and Berry's phase in graphene. Nature 2005, 438, 201–204.

139. Ozyilmaz, B.; Jarillo-Herrero, P.; Efetov, D.; Abanin, D.A.; Levitov, L.S.; Kim, P. Electronic transport and quantum hall effect in bipolar graphene p-n-p junctions. Phys. Rev. Lett. 2007, 99, 166804:1–166804:4.

140. Chen, J.H.; Jang, C.; Adam, S.; Fuhrer, M.S.; Williams, E.D.; Ishigami, M. Charged- impurity scattering in graphene. Nat. Phys. 2008, 4, 377–381.

141. Chen, J.H.; Jang, C.; Xiao, S.D.; Ishigami, M.; Fuhrer, M.S. Intrinsic and extrinsic performance limits of graphene devices on SiO2. Nat. Nanotechnol. 2008, 3, 206–209.

142. Lin, Y.M.; Avouris, P. Strong suppression of electrical noise in bilayer graphene nanodevices. Nano Lett. 2008, 8, 2119–2125.

143. Bolotin, K.I.; Sikes, K.J.; Jiang, Z.; Klima, M.; Fudenberg, G.; Hone, J.; Kim, P.; Stormer, H.L. Ultrahigh electron mobility in suspended graphene. Solid State Communications 2008, 146, 351–355.

144. Bolotin, K.I.; Sikes, K.J.; Hone, J.; Stormer, H.L.; Kim, P. Temperature dependent transport in suspended graphene. Phys. Rev. Lett. 2008, 101, 096802:1–096802:4.

145. Du, X.; Skachko, I.; Barker, A.; Andrei, E.Y. Approaching ballistic transport in suspended graphene.Nat. Nanotechnol. 2008, 3, 491–495.

146. Schedin, F.; Geim, A.K.; Morozov, S.V.; Hill, E.W.; Blake, P.; Katsnelson, M.I.; Novoselov, K.S. Detection of individual gas molecules adsorbed on graphene. Nat. Mater. 2007, 6, 652–655.

147. Gunlycke, D.; Lawler, H.M.; White, C.T. Room-temperature ballistic transport in narrow graphene strips. Phys. Rev. B 2007, 75, 085418:1–085418:5.

148. Ferrari, A.C. Raman spectroscopy of graphene and graphite: Disorder, electron–phonon coupling, doping and nonadiabatic effects. Solid State Commun. 2007, 143, 47–57.

149. Teague, M.L.; Lai, A.P.; Velasco, J.; Hughes, C.R.; Beyer, A.D.; Bockrath, M.W.; Lau, C.N.; Yeh, N.-C. Evidence for strain-induced local conductance modulations in single-layer graphene on SiO2. Nano Lett. 2009, 9, 2542–2546.

150. Guinea, F.; Horovitz, B.; Le Doussal, P. Gauge fields, ripples and wrinkles in graphene layers. Solid State Commun. 2009, 148, 1140–1143.

151. Han, M.Y.; Özyilmaz, B.; Zhang, Y.; Kim, P. Energy band-gap engineering of graphene nanoribbons.Phys. Rev. Lett. 2007, 98, 206805:1–206805:4.

152. Avouris, P.; Chen, Z.; Perebeinos, V. Carbon-based electronics. Nat. Nanotechnol. 2007, 2, 605–615.

153. Nakada, K.; Fujita, M.; Dresselhaus, G.; Dresselhaus, M.S. Edge state in graphene ribbons: Nanometer size effect and edge shape dependence. Phys. Rev. B 1996, 54, 17954–17961.

154. Son, Y.-W.; Cohen, M.L.; Louie, S.G. Energy gaps in graphene nanoribbons. Phys. Rev. Lett. 2006, 97, 216803–216804.

155. Kharche, N.; Zhou, Y.; O'Brien, K.P.; Kar, S.; Nayak, S.K. Effect of layer stacking on the electronic structure of graphene nanoribbons. ACS Nano 2011, 5, 6096–6101.

156. Chen, Z.; Lin, Y.-M.; Rooks, M.J.; Avouris, P. Graphene nanoribbon electronics. Phys. ELowDimens. Syst. Nanostruct. 2007, 40, 228–232.

157. Stankovich, S.; Piner, R.D.; Nguyen, S.T.; Ruoff, R.S. Synthesis and exfoliation of isocyanate-treated graphene oxide nanoplatelets. Carbon 2006, 44, 3342–3347.

158. Schniepp, H.C.; Li, J.-L.; McAllister, M.J.; Sai, H.; Herrera-Alonso, M.; Adamson, D.H.; Prud'homme, R.K.; Car, R.; Saville, D.A.; Aksay, I.A. Functionalized single graphene sheets derived from splitting graphite oxide. J. Phys. Chem. B 2006, 110, 8535–8539.

159. Cai, J.; Ruffleux, P.; Jaafar, R.; Bieri, M.; Braun, T.; Blankenburg, S.; Muoth, M.; Seitsonen, A.P.; Saleh, M.; Feng, X.; et al. Atomically precise bottom-up fabrication of graphene nanoribbons. Nature 2010,466, 470–473.

160. Li, X.; Wang, X.; Zhang, L.; Lee, S.; Dai, H. Chemically derived, ultrasmooth graphene nanoribbon semiconductors. Science 2008, 319, 1229–1232.

161. Dresselhaus, M.S.; Dresselhaus, G. Intercalation compounds of graphite. Adv. Phys. 2002, 51, 1–186.

162. Jiao, L.; Wang, X.; Diankov, G.; Wang, H.; Dai, H. Facile synthesis of high-quality graphene nanoribbons. Nat. Nanotechnol. 2010, 5, 321–325.

163. Naeemi, A.; Meindl, J.D. Performance Benchmarking for Graphene Nanoribbon, Carbon Nanotube, and Cu Interconnects. In Proceedings of International Interconnect Technology Conference, Burlingame, CA USA, 14 June 2008; pp. 183–185.

164. Balandin, A.A.; Ghosh, S.; Bao, W.; Calizo, I.; Teweldebrhan, D.; Miao, F.; Lau, C.N. Superior thermal conductivity of single-layer graphene. Nano Lett. 2008, 8, 902–907.

165. Ouyang, Y.; Wang, X.; Dai, H.; Guo, J. Carrier scattering in graphene nanoribbon field-effect transistors. Appl. Phys. Lett. 2008, 92, 243124:1–243124:3.

166. Wang, H.M.; Wu, Y.H.; Ni, Z.H.; Shen, Z.X. Electronic transport and layer engineering in multilayer graphene structures. Appl. Phys. Lett. 2008, 92, 053504:1–053504:3.

167. Lu, Y.H.; Chen, W.; Feng, Y.P.; He, P.M. Tuning the electronic structure of graphene by an organic molecule. J. Phys. Chem. B 2009, 113, 2–5.

168. Wang, X.; Li, X.; Zhang, L.; Yoon, Y.; Weber, P.K.; Wang, H.; Guo, J.; Dai, H. N-doping of graphene through electrothermal reactions with ammonia. Science 2009, 324, 768–771.

169. Wei, D.; Liu, Y.; Wang, Y.; Zhang, H.; Huang, L.; Yu, G. Synthesis of N-doped graphene by chemical vapor deposition and its electrical properties. Nano Lett. 2009, 9, 1752–1758.

170. Wehling, T.O.; Novoselov, K.S.; Morozov, S.V.; Vdovin, E.E.; Katsnelson, M.I.; Geim, A.K.; Lichtenstein, A.I. Molecular doping of graphene. Nano Lett. 2008, 8, 173–177.

171. Lee, E.J.H.; Balasubramanian, K.; Weitz, R.T.; Burghard, M.; Kern, K. Contact and edge effects in graphene devices. Nat. Nanotechnol. 2008, 3, 486–490.

172. Faugeras, C.; Nerriere, A.; Potemski, M.; Mahmood, A.; Dujardin, E.; Berger, C.; de Heer, W.A. Few-layer graphene on SiC, pyrolitic graphite, and graphene: A Raman scattering study. Appl. Phys. Lett.2008, 92, 011914:1–011914:4.

173. Cervantes-Sodi, F.; Csanyi, G.; Piscanec, S.; Ferrari, A.C. Edge-functionalized and substitutionally doped graphene nanoribbons: Electronic and spin properties. Phys. Rev. B 2008, 77, 165427:1–165427:13.

12

ASSEMBLING SINGLE GOLD NANORODS INTO LARGE-SCALE HIGHLY ALIGNED NANOARRAYS VIA VACUUM-ENHANCED CAPILLARITY

Jiaojiao Wang[1,2], Min Li[2], Bochong Tang[2], Peng Xie[2], Lei Ma[2], Zhongbo Hu[1], Yuliang Zhao[2] and Zhongqing Wei[2]

[1]College of Materials Science and Opto-electronic Technology, University of Chinese Academy of Sciences, Yuquan Rd. 19A, Beijing 100049, China.

[2]CAS Key Laboratory for Biomedical Effects of Nanomaterials and Nanosafety, Institute of High Energy Physics, Chinese Academy of Sciences, Yuquan Rd. 19B, Beijing 100049, China

ABSTRACT

We report a simple, straightforward, and efficient approach to assemble single gold nanorods (AuNRs) into highly aligned arrays, via a unique vacuum-enhanced capillarity. The assembled AuNR arrays demonstrate both an excellently unidirectional ordering and a wonderful single-rod resolution. The key role of vacuum in this approach enables high-aspect-ratio (10 to 22) AuNR alignment and efficiently facilitates large-area alignment. Further investigation of one- and two-dimensional AuNR arrays would undoubtedly be beneficial to their potential applications.

Keywords: Gold Nanorods Assembly Capillary Atomic force microscopy Nanoarray

BACKGROUND

Controlled assembly and alignment of one-dimensional (1D) nanostructures such as nanorods, nanotubes, and nanowires are essential for their integration and applications in many macroscopic devices for nanoelectronics, [1] sensing, [2, 3, 4], and plasmonics [5, 6]. Although 1D nanostructures such as gold nanoarrays can be fabricated by conventional top-down lithography (e.g., electron beam lithography), the process is time-consuming and the gold involved is polycrystalline, which may ultimately degrade the performance of devices. Recent wet chemical approaches have made it possible to synthesize various anisotropic gold nanostructures including dimeric gold nanorod (AuNR) junctions [4, 5, 7]. Meanwhile, several novel bottom-up strategies, such as capillary-driven assembly, [8, 9, 10] chemically template-directed assembly [11, 12, 13], spontaneous self-assembly [14], surface amidation assembly [15], and polymer-based assembly [16, 17, 18], have been developed to assemble spherical gold colloid nanoparticles [7] and anisotropic pentahedrally twinned AuNRs into ordered gold nanostructures [11, 15, 19]. However, the manipulation of anisotropic AuNRs to form highly aligned and ordered nanoarrays still remains significantly challenging. Herein, we employ a strong capillarity enhanced by a vacuum created from a syringe to assemble AuNRs into nanoarray structures. The assembled linear nanoarrays are not only single-rod resolved but also extremely well aligned. Compared with the spontaneous capillary assembly, this approach is simple, quick, and clean (does not involve any other chemicals such as polymers) and can be highly efficient in large-area assembly of AuNRs with high aspect ratios (ARs) (approximately 10 to 22). The vacuum-enhanced capillary force effectively promotes the formation of highly aligned AuNR arrays. This work is an extended study of our previous work on gold nanorods [20].

METHODS

Figure 1 outlines our procedure depicting the assembly of AuNRs using a syringe as a simple pump to generate a vacuum-enhanced capillary force. Briefly, a piece of cleaned Si (100) substrate was sandwiched in between two pieces of semi-cylindrical poly(dimethylsiloxane) (PDMS) stamps, one of which has linear periodic nanochannels (a) molded from a DVD template. Note that in the present work, we used two complementary DVD templates (referred to as template 1 and template 2 hereafter, see Additional file 1: Figure S1 and Additional file 1: Figure S2 for details) to fabricate two types of PDMS stamps with nanochannels. A piece of tape was then wound tightly around the two pieces of PDMS stamps to clamp them together and hold the silicon substrate in position. Following dropping a few drops (approximately 5 μl) of highly purified AuNR dispersion (approximately 0.05 mg/ml) on the protruding portion of the substrate surface, the entire assembly was then carefully squeezed into the cylindrical tube of a syringe after the plunger had been pulled away from the syringe body. At this time, the nanochannels may deform due to the pressure imposed by the internal wall of the tube, giving rise to the formation of nanopore canals (b) with a diameter of ca. 20 nm. The plunger was pushed back into the tube, followed by sealing the open end of the syringe with a piece of tape. Thus, a closed system was created. Pulling the plunger produced a vacuum inside the tube between the ends of the plunger and stamps. Due to the vacuum-enhanced pressure difference between both ends of the stamps along inside the tube, the flow of the AuNR dispersion occurred spontaneously and rapidly to fill up (d) the nanopore canals. After the whole assembly was then left for several hours for the water to evaporate completely, the PDMS stamps were gently removed, leaving patterned AuNR arrays on the silicon surface (e). A crossed AuNR array can be formed by repeating steps from (a) to (e) to generate another linear array perpendicular to the existing linear array (f-h).

RESULTS AND DISCUSSION

We first synthesized AuNRs following the procedure described by Murphy et al. [21] and then purified them following the procedure reported by Zubarev and Khanal [22] [see Additional file 1: Figure S3]. In Figure 2, the aqueous dispersion of thus-prepared AuNRs shows the Tyndall effect, revealing its colloidal nature. Typically, its UV-Vis-NIR spectrum shows one transversal surface plasma peak at about 498 nm and one longitudinal peak at about 1,720 nm. The redshifted maximum of the longitudinal plasmon absorption and the blueshifted maximum of the transverse plasmon absorption are due to the high AR (ca. 10 to 22, see below) of the AuNRs. Figure 2A shows a representative SEM image of AuNRs distributed on the surface of the n-type Si (100) substrate which was previously functionalized with (3-mercaptopropyl)triethoxysilane (MPTES) [see Additional file 1]. Obviously, the AuNRs are highly purified (approximately 100%), size uniform, and well isolated (no aggregates). Larger-area observations

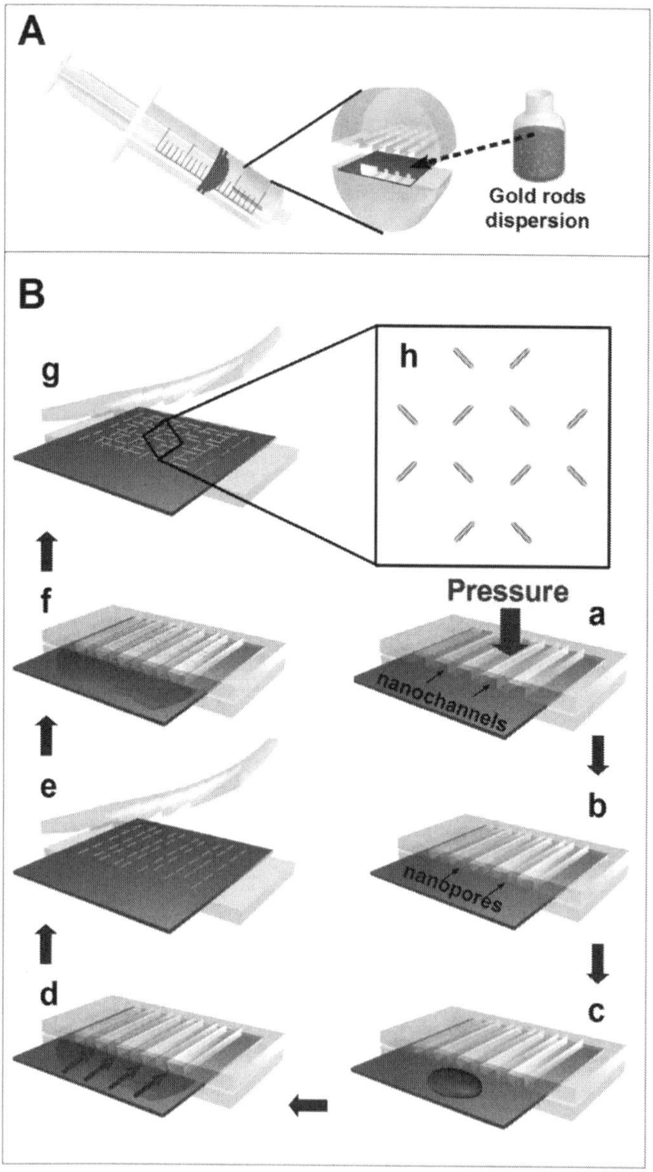

Figure 1. Schematic drawing depicting the formation of AuNR arrays driven by a vacuum-enhanced capillary force (A, B). See text for details.

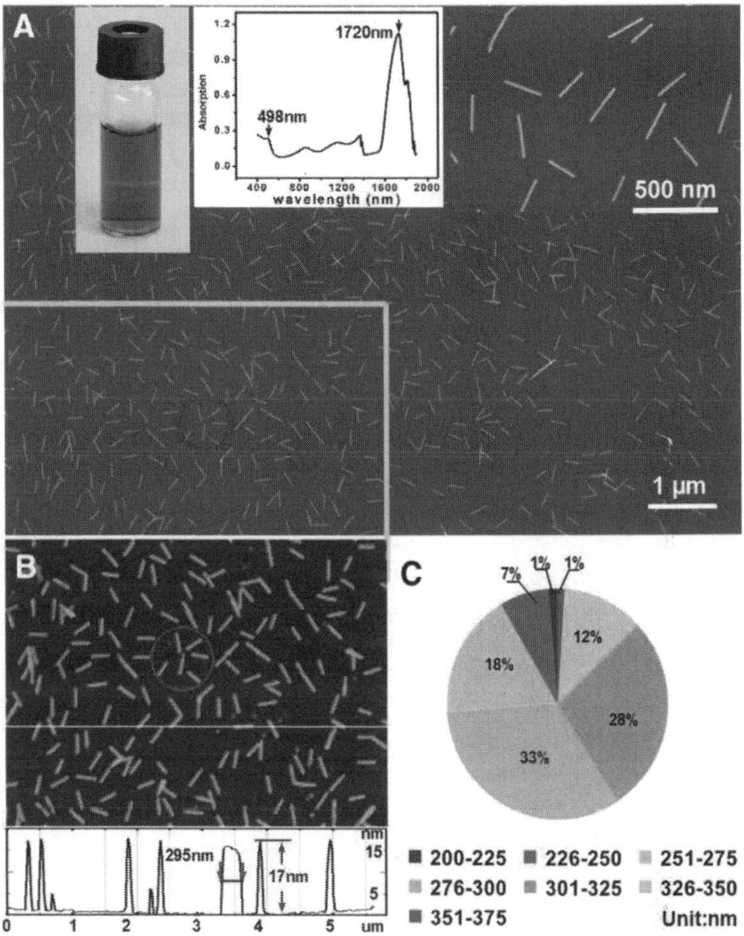

Figure 2. Highly purified AuNRs used to fabricate AuNR arrays. (A) SEM image of AuNRs uniformly distributed on a MPTMS-modified silicon surface. Insets are a digital picture of AuNR dispersion (approximately 0.1 mg/ml) showing the Tyndall effect, UV-Vis spectrum, and close-up view of AuNRs, respectively. **(B)** *In situ* AFM image of AuNRs captured in the same area as outlined area in the SEM image. The circled items in the two images are used as markers to demonstrate the same areas. **(C)**Statistical distribution of lengths of 722 AuNRs.

(>55 μm, Additional file 1: Figure S4) also show the same situation. We measured the size of 722 single AuNRs (not all of them are shown) in the AFM image (Figure 2B), which was captured in the same area as outlined in the SEM image (Figure 2A). AFM measurements indicate that all of the AuNRs possess a length of over 200 nm with a diameter of ca. 17 to 20 nm. Shown in Figure 2C is a pie chart depicting the statistics analysis of lengths for measured AuNRs. The average high AR for AuNRs ranges from ca. 10 to 22.

The SEM images presented in Figure 3 show highly aligned AuNR arrays assembled on silicon surfaces with the aid of a vacuum-enhanced capillary force. It is obvious that the AuNRs were aligned extremely well (Figure 3B). The spacing in the periodic array structure is determined exactly by the dimension of the PDMS stamp used. In Figure 3A, while some AuNRs succeeded in entering nanopores under vacuum-enhanced capillary and then aligning, most of the single AuNRs accumulated vastly on the entry positions, giving rise to AuNR aggregations. The schematic diagram shown in Figure 3C is the mechanism we proposed to illustrate the entire process of forming aligned arrays. Under the pressure imposed by the internal wall of the syringe tube, the PDMS stamp (b), molded from template 1 (a), sagged down to generate nanopore canals. The size of the resulting canals matched the diameter (approximately 20 nm) of AuNRs so well that only one single AuNR could enter the canal at a time, leading to a highly aligned linear array at a single-AuNR level. Large-area aligned AuNR arrays can be seen in Additional file 1: Figure S5. The complete match in Figure 3B between the model (inset) and the alternatively bright and dark press marks left by the PDMS stamps strongly supports our proposed mechanism.

Subtle differences (e.g., the elastic modulus of stamps, the pressure on stamps) in practical operations can result in different patterning. Shown in Figure 4 are two variants of the AuNR arrays presented in Figure 3. In Figure 4A, a relatively soft PDMS stamp cast from template 1 [see Additional file 1] was used to assemble AuNRs. The array generated has an equally spaced array structure (a period of approximately 400 nm). We speculate that when the soft stamp is in contact with the silicon surface, the raised areas of the stamp deform laterally as a result of the pressure, causing the size increase laterally (from approximately 300 to 400 nm) in contact areas and size decrease laterally (from approximately 500 to 400 nm) in noncontact areas (channel areas). The schematic drawing below the images depicts this process of size changes. Thus, by tuning the elastic modulus of the stamp or/and the pressure imposed on the stamp, linear AuNR arrays with varied spacing can be obtained. Another example shown in Figure 4B demonstrates the effect of less pressure applied to the stamp on the final patterning. In Figure 4B, the stamp (molded from template 2) exerted less pressure and the nanochannels failed to turn into nanopores. As a result, bundles of AuNRs, rather than single ones, entered the nanochannels, giving rise to an array composed of bundles of AuNRs. An additional image of the bundled AuNR array can be seen in Additional file 1: Figure S6.

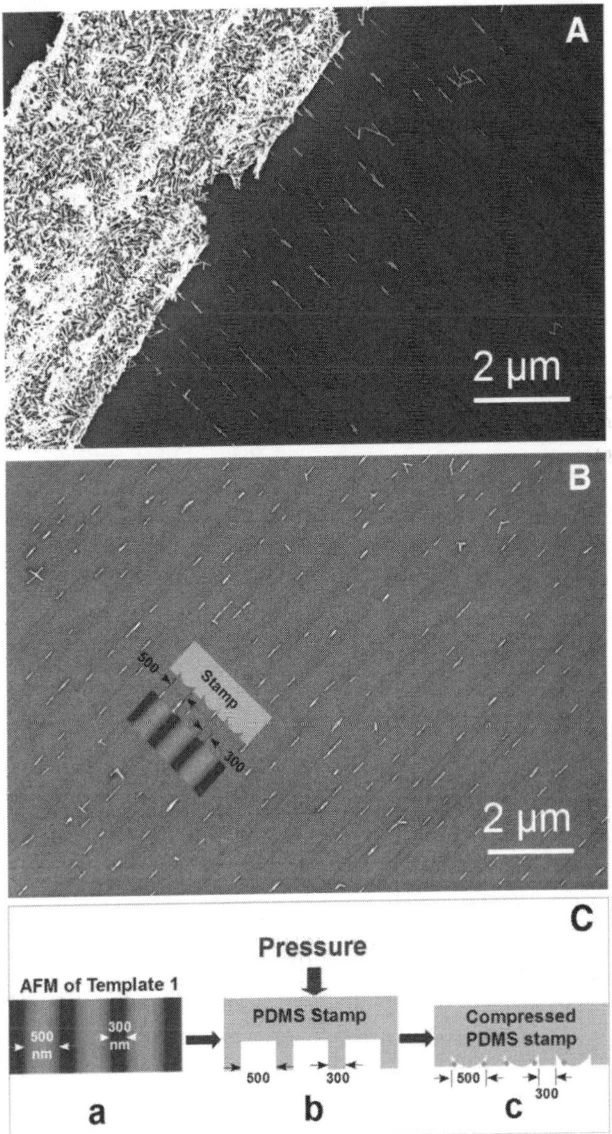

Figure 3. Highly aligned AuNRs generated via the procedure described in Figure 1. **(A)** Aggregated and single AuNRs before and after entering nanopores, respectively. **(B)** Representative AuNR array showing a resolution of single AuNRs. **(C)** Model used to illustrate the process of array formation.

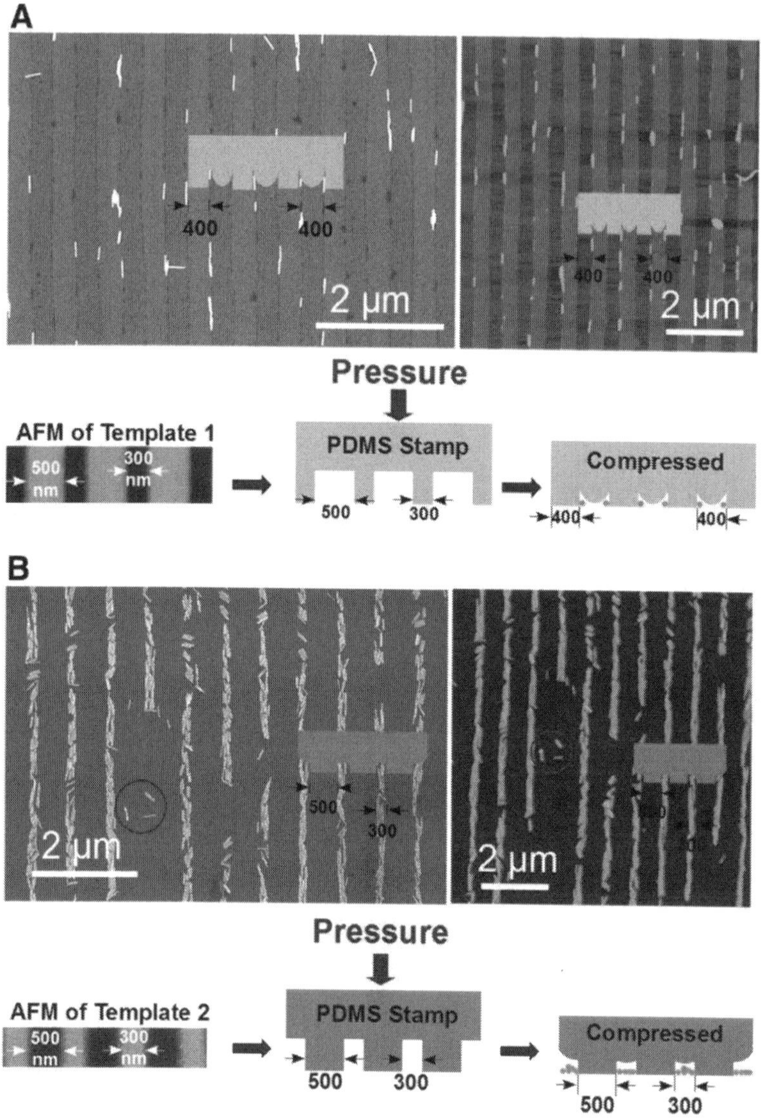

Figure 4. The structure of AuNR arrays can be tuned by adjusting parameters related to PDMS stamps. (A) SEM and AFM images of an equally spaced array assembled using a relatively soft stamp. Shown below the images is a model to illustrate the process. **(B)** SEM and corresponding *in situ* AFM images of an array (composed of bundles of AuNRs) assembled by applying smaller pressure to the stamp. The blue circled items in the two images are used as markers to demonstrate the same areas. Shown below the images is a model to illustrate the process.

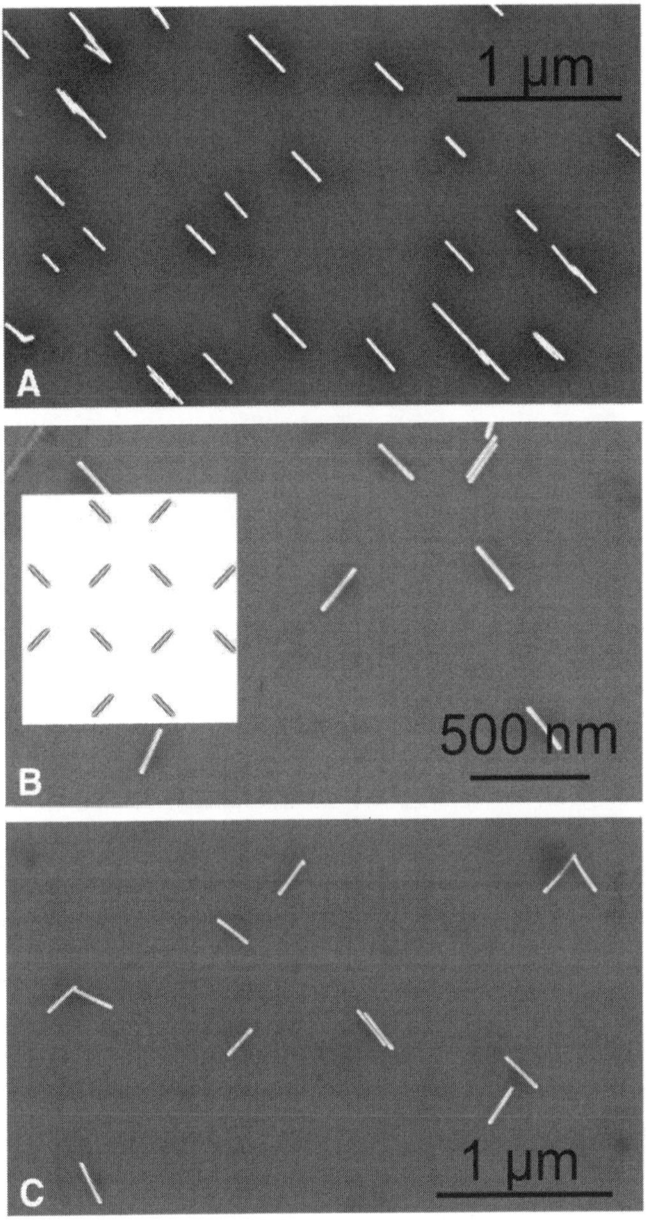

Figure 5. AuNR arrays with network structures were generated using double assembling. The linear array was rotated by 90° before making the second assembling. **(A)** The linear array generated from the first assembling. **(B, C)** Network arrays generated after the second assembling.

Normal capillary force was demonstrated to be useful for patterning nanomaterials including spherical nanoparticles [8, 23]. Our control experiments show, however, that normal capillary force failed to drive high-AR (ca. 10 to 22 in this work) AuNRs into the nanopores, and arrays of AuNRs could therefore not be obtained. It is reasonable to assume that driving a higher-AR AuNR needs much higher force than driving a spherical gold nanoparticle or a lower-AR AuNR with a comparable diameter. This is due to the fact that a higher-AR AuNR is normally much heavier than a lower-AR AuNR or a spherical nanoparticle. Vacuum created by a simple syringe in the present work provides an enhanced capillary force, making the assembly of high-AR AuNRs successful and efficient.Variants of linear AuNR arrays could be obtained by choosing diverse templates with different dimensions or/and relief structures or by multiple linear assemblies with different orientations on a substrate. Shown in Figure 5 is a proof-of-concept example demonstrating the feasibility to form a two-dimensional AuNR array. After a linear AuNR array (Figure 5A) was produced using the patterning procedure described in Figure 1, we rotated the patterned substrate 90° in its plane and then placed another linear array of AuNRs by carrying out the same patterning procedure. The resultant array displays a two-dimensional network structure (Figure 5B,C), as illustrated in Figure 1B-h.

CONCLUSIONS

We have developed a simple, inexpensive, and efficient procedure that is based on a vacuum-enhanced capillary force for assembling anisotropic AuNRs into AuNR arrays. The AuNR arrays generated are not only single-rod resolved but also highly aligned and ordered. The created vacuum enables fast alignment of high-AR AuNRs on a large scale. Further work is needed to investigate the impact of various parameters on assemblies and to refine the process of forming two-dimensional AuNR nanoarrays as demonstrated in the proof-of-concept example. The approach presented here could be extended to assemble other anisotropic nanostructures including other nanorods/nanowires, carbon nanotubes, and DNA, and the nanoarrays fabricated could find potential applications in nanoelectronics, nanoplasmonics [5, 6], SERS sensing [6], and chemical detection [24].

Acknowledgements

This work was supported by the 100 Talents Program of the Chinese Academy of Sciences (Y1291720S3) and the National Basic Research Program of the Ministry of Science and Technology of China (2011CB933101).

COMPETING INTERESTS

The authors declare that they have no competing interests.

AUTHORS' CONTRIBUTIONS

JW synthesized, assembled, and characterized the gold nanorods and participated in the interpretation of experimental results and in the manuscript drafting. ZW conceived, designed, and guided the research and drafted the manuscript. ML, BT, PX, LM, ZH, and YZ provided assistance in the experiments and discussed the data. All authors read and approved the final manuscript.

REFERENCES

1. Rao SG, Huang L, Setyawan W, Hong S: Nanotube electronics: large-scale assembly of carbon nanotubes.*Nature* 2003, 425: 36–37.

2. Li CZ, Male KB, Hrapovic S, Luong JHT: Fluorescence properties of gold nanorods and their application for DNA biosensing. *Chem Commun* 2005, 43: 3924–3926.

3. Vigderman L, Khanal BP, Zubarev ER: Functional gold nanorods: synthesis, self-assembly, and sensing applications. *Adv Mater* 2012, 24: 4811–4841.

4. Vigderman L, Zubarev ER: Starfruit-shaped gold nanorods and nanowires: synthesis and SERS characterization. *Langmuir* 2012, 28: 9034–9040.

5. Wild B, Cao L, Sun Y, Khannl BP, Zubarev ER, Gray SK, Scherer NF, Pelton M: Aligned growth of gold nanorods in PMMA channels: parallel preparation of nanogaps. *ACS Nano* 2012, 6: 472.

6. Grzelczak M, Liz-Marzán LM: Colloidal nanoplasmonics: from building blocks to sensing devices. *Langmuir* 2013, 29: 4652–4663.

7. Jain T, Tang Q, Bjørnholm T, Nørgaard K: Wet chemical synthesis of soluble gold nanogaps. *Acc Chem Res* 2013, 47: 2–11.

8. Cui Y, Björk MT, Liddle JA, Sönnichsen C, Boussert B, Alivisatos AP: Integration of colloidal nanocrystals into lithographically patterned devices. *Nano Lett* 2004, 4: 1093–1098.

9. Rey A, Billardon G, Lortscher E, Moth-Poulsen K, Stuhr-Hansen N, Wolf H, Bjornholm T, Stemmer A, Riel H: Deterministic assembly of linear gold nanorod chains as a platform for nanoscale applications. *Nanoscale* 2013, 5: 8680–8688.

10. Kuemin C, Stutz R, Spencer ND, Wolf H: Precise placement of gold nanorods by capillary assembly. *Langmuir* 2011, 27: 6305–6310.

11. Liu S, Tok JBH, Locklin J, Bao Z: Assembly and alignment of metallic nanorods on surfaces with patterned wettability. *Small* 2006, 2: 1448–1453.

12. Nepal D, Onses MS, Park K, Jespersen M, Thode CJ, Nealey PF, Vaia RA: Control over position, orientation, and spacing of arrays of gold nanorods using chemically nanopatterned surfaces and tailored particle–particle–surface interactions. *ACS Nano* 2012, 6: 5693–5701.

13. Holzner F, Kuemin C, Paul P, Hedrick JL, Wolf H, Spencer ND, Duerig U, Knoll AW: Directed placement of gold nanorods using a removable template for guided assembly. *Nano Lett* 2011, 11: 3957–3962.

14. Nikoobakht B, Wang ZL, El-Sayed MA: Self-assembly of gold nanorods. *J Phys Chem B* 2000, 104: 8635–8640.

15. Mieszawska AJ, Slawinski GW, Zamborini FP: Directing the growth of highly aligned gold nanorods through a surface chemical amidation reaction. *J Am Chem Soc* 2006, 128: 5622–5623.

16. Jain T, Lara-Avila S, Kervennic YV, Moth-Poulsen K, Nørgaard K, Kubatkin S, Bjørnholm T: Aligned growth of gold nanorods in

PMMA channels: parallel preparation of nanogaps. *ACS Nano* 2012, 6: 3861–3867.

17. Murphy CJ, Orendorff CJ: Alignment of gold nanorods in polymer composites and on polymer surfaces.*Adv Mater* 2005, 17: 2173–2177.

18. Hore MJA, Composto RJ: Functional polymer nanocomposites enhanced by nanorods. *Macromolecules*2013, 47: 875–887.

19. Khanal BP, Zubarev ER: Rings of nanorods. *Angew Chem Int Ed* 2007, 46: 2195–2198.

20. Wei Z, Qi H, Li M, Tang B, Zhang Z, Han R, Wang J, Zhao Y: Watching single gold nanorods grow. *Small*2012, 8: 1331–1335.

21. Jana NR, Gearheart L, Murphy CJ: Wet chemical synthesis of high aspect ratio cylindrical gold nanorods. *J Phys Chem B* 2001, 105: 4065–4067.

22. Khanal BP, Zubarev ER: Purification of high aspect ratio gold nanorods: complete removal of platelets. *J Am Chem Soc* 2008, 130: 12634–12635.

23. Suh KY, Kim YS, Lee HH: Capillary force lithography. *Adv Mater* 2001, 13: 1386–1389.

24. Ma W, Kuang H, Xu L, Ding L, Xu C, Wang L, Kotov NA: Attomolar DNA detection with chiral nanorod assemblies. *Nature Commun* 2013, 4: 2689–3695.

13

FABRICATION AND ELECTRICAL CHARACTERIZATION OF HIGHLY ORDERED COPPER NANOWIRES

Gurmeet Singh Lotey[1] • Sanjeev Kumar[2]• N. K. Verma[1]

[1].Nano Research Lab, School of Physics and Materials ScienceThapar UniversityPatialaIndia

[2].University College of Engineering, Punjabi UniversityPatialaIndia

ABSTRACT

The template-assisted electrodeposition technique has been employed to synthesize highly ordered: uniformly dense, well-aligned, parallel and homogeneous copper nanowires. Their morphological studies have been carried out using the scanning electron microscopy and transmission electron microscopy. The X-ray diffraction study exhibits cubic structure of the nanowires and their preferred orientation along the direction [111]. Their elemental composition has been done by energy dispersive X-ray analysis. The photoluminiscence spectra of copper nanowires show two excitation peaks at 209 and 268 nm; both these absorption pathways yield fluorescence at 296 nm. Their UV–Vis absorption spectra have been found to give a prominent peak at 570 nm. The current–voltage characteristics of the nanowires reveal their non-linear behavior. The impedance spectroscopy has also been carried out and it shows an increase in their impedance at higher frequencies.

Keywords: Electrodeposition Nanowires Nonlinear behavior Impedance I

INTRODUCTION

One-dimensional (1D) nanostructures such as nanotubes, nanowires, nanobelts have been extensively studied not only for their novel physical, chemical, electronic, magnetic, and electrical properties but also for their prospective applications in future-generation nanodevices. Among the various hierarchical structures, 1D nanostructure arrays, perpendicular to the substrate, are of significant interest. The 1D nanostructures are ideal building blocks for nanoelectronics (Liu and Bando 2003) since, in the architecture of nanodevices, they can act both as interconnects and device components. The miniaturization of basic device elements, using different types of junctions, leads to the vast success in microelectronics industry. The primary goal of nanoelectronics industry is to achieve similar functionality at nanoscale. However, the basic problem arises when the essential device elements are joined together by 1D nanostructures to form multi-terminal nanodevices eventually leading to complex circuits. Thus, the study on highly ordered metal nanowires is of great motivation (Zhang et al. 2005; Zach et al. 2000). The metallic nanowires have potential applications in a wide range of advanced fields (Zhang et al. 2005; Zach et al. 2000; Toma et al.2010; Mehrez and Guo 2004; Wildoer et al. 1998). Due to accessibility and exceptional properties such as good strength, admirable malleability, and superior corrosion resistance, copper is the third most broadly used commercial metal after iron and aluminum. Furthermore, copper has outstanding electrical and thermal conductivity second to silver. Therefore, copper nanowires are of particular importance. With the fast reduction in size of electronic devices, copper nanowires play a vital role as interconnects in nanoelectronics and optoelectronic devices. The number of methods used for the fabrication of such nanowires, includes vapor deposition (Liu and Bando 2003; Zhang et al. 2005), reverse micelle system (Lisiecki et al.1996), electrochemical synthesis (Liu et al. 2008; Bicelli et al. 2008; Zhang and Wong 2009), ion-beam sputtering (Toma et al. 2010), solution-phase synthesis (Wang et al. 2008), etc. (Choi and Park 2004; Zhang et al. 2008; Duan et al. 2010; Monson and Woolley 2003; Lai and Riley2008; Qin et al. 2007). An ordinary fabrication method for nanowires is based on the electrodeposition of metals into pores of controlled geometry. Nanostructures synthesized by template-assisted technique, using anodic alumina membrane

(AAM) (Furneaux et al. 1989; Li et al. 1998), are quite attractive due to these templates' remarkable hardness, uniform pore size, high pore density together with relatively low cost. The electrodeposition method is influenced by several parameters, e.g., temperature, pH, overvoltage, and composition of the electrolyte, which have been studied for different systems (Gao et al. 2002; I.U. Schuchert et al. 2003; Pan et al. 2005a, 2005b; Pan et al. 2005a, 2005b).

This paper presents synthesis of highly ordered copper nanowires using template-assisted electrodeposition technique. The electrical characterization viz. current–voltage (I–V) characteristics and the impedance spectroscopy of the nanowires are also being studied. The nanowires showed a unique behaviour; they did not obey Ohm's law rather the charge carriers were found to tunnel through a potential barrier; also an increase in impedance was noticed at high frequencies.

EXPERIMENTAL

The commercially available templates (Anodisc 25, Whatman, UK) with average pore diameter, 100 nm, and pore density, 10^9 cm^{-2} have been used for the fabrication of copper nanowires. The chemicals and reagents used were of analytical grade. The de-ionized water was used for preparing solutions. The electrodeposition process has been carried out in a typical three-electrode electrochemical cell using Potentiostat (Gamry Reference 600) with a platinum electrode as counter electrode, $Ag/AgCl_2$ as reference electrode, and the AAM coated with Ag as working electrode. 0.1 M $CuSO_4.5H_2O$ aqueous solution was used as electrolyte, and the pH of this solution was adjusted to be 2 using 0.1 M H_2SO_4. A high concentration of $CuSO_4$ is required to supply a sufficiently large number of copper ions (Cu^{2+}) inside the pores of AAM during the electrodeposition process. H_2SO_4 increases the conductivity of the solution and lowers the cathode overvoltage Using a cyclic voltametry curve (Fig. 1), the reduction potential, required for depositing copper, was found to be 1.2 V. By applying this low voltage, side reactions such as evolution of hydrogen bubbles were avoided.

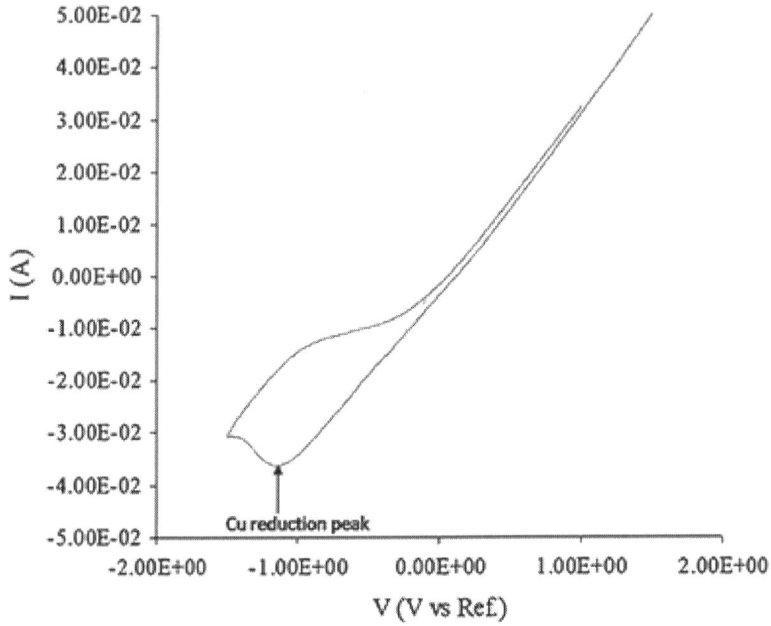

Figure 1. The cyclic voltammetry curve for copper electrodeposition in anodic alumina membrane

CHARACTERIZATION AND MEASUREMENTS

The crystallographic study of copper nanowires embedded in AAM as host was performed using the X-ray diffractometer (PANalytical X'Pert PRO MRD ML) with CuKα (λ = 1.5418 Å) radiation operated at 45 kV and 40 mA. The high intense beam was focused over a small area (10 mm^2) of the sample, and gonio scan was recorded for 2θ values from 10° to 90°. The KEITHLEY source meter, 4200-SCS, attached to ZYVEX S100 Nanomanipulator, was used for collective I–V measurements. The impedance spectroscopy was carried out using Gamry Reference 600 potentiostat and an especially designed Faraday cage at frequencies from 0.2 Hz to 1 MHz. For scanning electron microscopy (SEM) characterization, the prepared sample of copper nanowires embedded in the AAM was put in a solution of NaOH to dissolve the alumina membrane. A very small concentration (0.01 M) of NaOH was used to prevent damage of the nanowires. The nanowires were then washed several times with de-ionized water followed by ethanol. A small drop of the above solution, containing the nanowires, was taken for the SEM and TEM characterizations. For SEM analysis, the dried sample of copper nanowires was coated with gold–palladium alloy using JEOL, JFC sputter coater and then examined under JEOL, JSM-6510 L scanning electron microscope. The energy dispersive X-ray spectroscopy (EDAX) was used for elemental composition analysis using Noran System Six attached to SEM. The transmission electron microscopy (TEM) measurements were taken with a Hitachi H-7500 system operating at 110 kV. The optical characterization of copper nanowires has been done by UV–Vis spectroscopy using UV-Thermoevolution spectrometer. The photoluminescence (PL) study has been carried out using Cary Varian fluorospectrophotometer with xenon lamp as an excitation source.

RESULTS AND DISCUSSION

Growth and morphology study

Figure 2a and b show the SEM micrographs respectively for the cross-sectional and lateral view of the nanowires showing their dense growth. They are found to be uniformly dense, homogeneous, parallel, and well aligned. The nanowires are found to be of size, 100 nm, and length, 12 μm as confirmed by TEM (Fig. 3). The aspect ratio of the nanowires is about 120.

The nucleation and growth inside the nanochannels of AAM result in the formation of copper nanowires; their growth is based on the reduction of Cu^{2+} ions by electrodeposition as shown below:

$$CuSO_4 \rightarrow Cu^{2+} + SO^{4-}$$

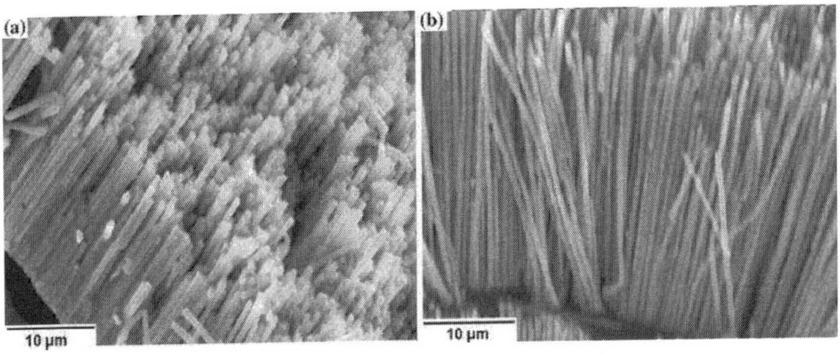

Figure 2. The SEM micrographs of copper nanowires

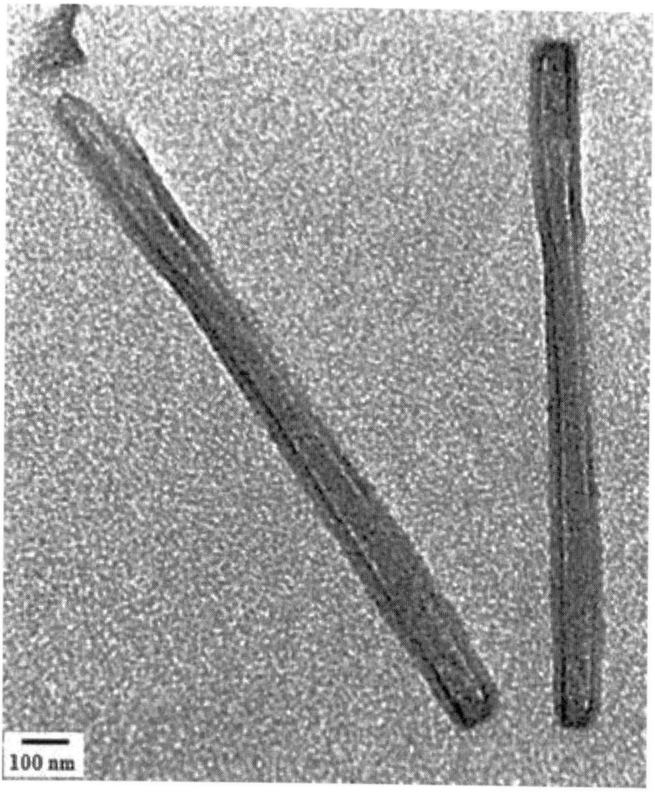

Figure 3. The transmission electron microscopy (TEM) images of copper nanowires

The Cu^{2+} ions are introduced into the nanochannels. On applying the electric field, they get reduced on the surface of the working electrode as follows:

$$Cu^{2+} \rightarrow Cu + 2e^{-}$$

First, the Cu nuclei are formed at the base of the template, and then the copper nanowires grow along the pores because of the trapping effect of the template. Their growth is controlled by the parameters such as applied potential, concentration of ions, flow rate of the electroactive species and pH value of the solution.

Structural study

The X-ray diffraction (XRD) analysis of the copper nanowires, embedded in AAM, has been carried out to study crystal structure (Fig. 4). It shows the presence of the four reflection peaks attributed to planes, (1 1 1), (2 0 0), (2 2 0), (3 1 1), confirming copper nanowires with the face-centered cubic crystal structure. From Fig. 4, it is also clear that the growth of copper nanowires is preferably along the direction, [111]. The reflection peaks observed for silver may be due to some silver paste still left at the AAM. The EDAX spectrum of copper nanowires (Fig. 5) exhibits the peaks of copper along with those of gold; as the latter is required for SEM. Some small peaks due to environmental carbon and oxygen are also observed.

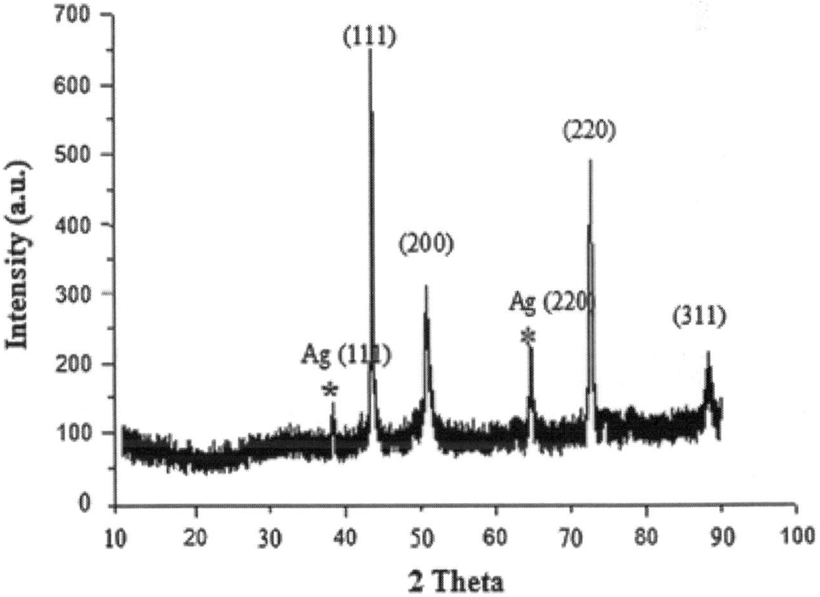

Figure 4. The XRD pattern of copper nanowires

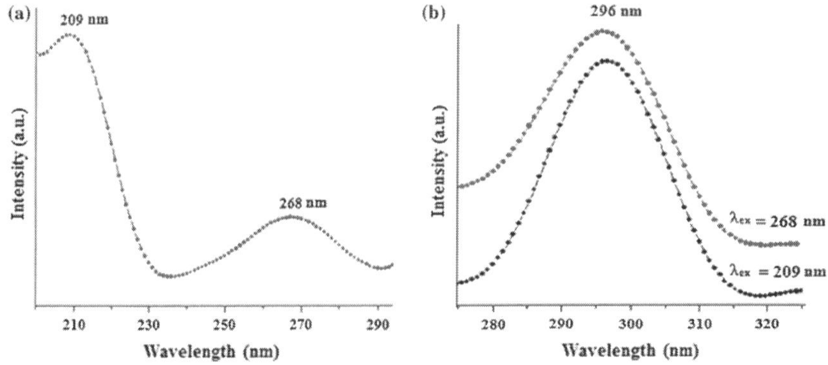

Figure 5. **a** The fluorescence excitation spectra of copper nanowires and **b** fluorescence emission spectra at excitation wavelengths 209 and 268 nm

Optical study

Photoluminescence study

Two peaks at 268 and 209 nm, corresponding to maximum transition probability/resonant absorption, have been observed (Fig. 6a) in the fluorescence excitation spectra; these peaks correspond, respectively, to the transition between the ground state (E_0) and the first excited state (E_1), and the transition between the ground state and the second excited state (E_2). Both of these absorption pathways have been found to yield fluorescence at 296 nm. Both the excitation wavelengths, 209 and 268 nm, yield the fluorescence emission at 296 nm (Fig. 6b); this emission peak has been found to remain fixed whatever the excitation wavelengths. The fluorescence from noble metals has been attributed to transitions of electrons in the conduction band below the Fermi level to the holes in the *d* bands (Mooradian 1968; Boyd et al. 1986). The peak observed at 296 nm in fluorescence spectra is due to transitions from the excited states to *d*-levels of the copper nanowires (Siwach and Sen 2008). However, the intensity of fluorescence peaks varies with change in λ_{ex} (Fig. 6b) because of the change in transition probability.

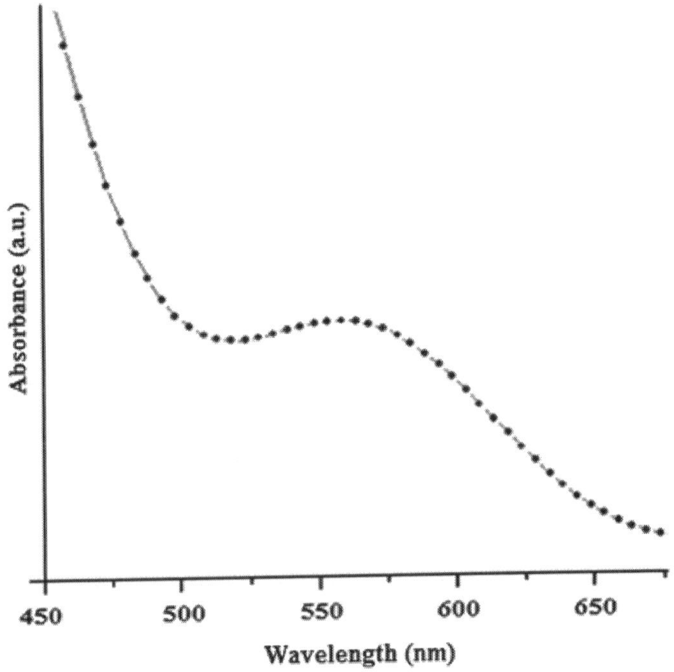

Figure 6. UV–vis absorption spectrum of copper nanowires

UV–visible analysis

The metal nanoparticles exhibit absorption bands or broad regions of absorption in the UV–visible range due to the excitation of surface plasmon resonances (SPR) or interband transitions; these SPR are characteristic properties for the metallic nature of particles. The prominent peak observed at 570 nm, in the visible wavelength, is due to the absorption of surface plasmon (Fig. 7). This broadband absorption, in the copper nanowires, is attributed to SPR and its tail at wavelength 510 nm, to *d–sp* interband transitions. The bandwidth of the resonance increases due to the decrease in size of the particles as well as the enhancement in the scattering of electrons at the surface of nanowires.

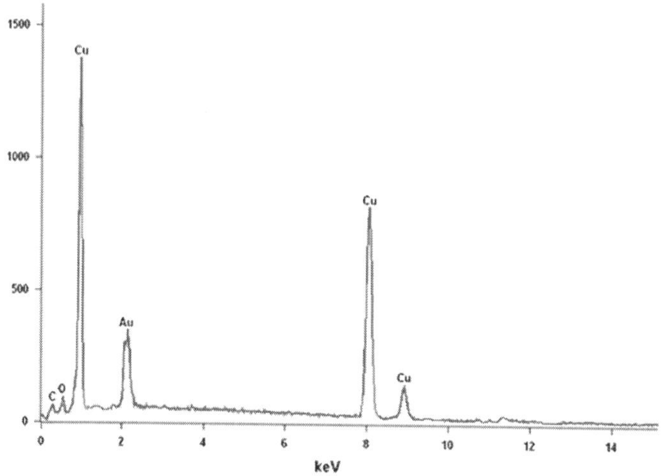

Figure 7. The EDAX spectrum of copper nanowires

Electrical and transport properties

Current voltage characteristics

Figure 8a shows the schematic set up for the measurement of collective I–V characteristics of copper nanowires. Figure 8b displays their non-linear and symmetric characteristics, with respect to zero bias. This exhibits a very unique behaviour of the nanowires that they do not obey the Ohm's law rather the electrons, the charge carriers, are being tunneled through a potential barrier. The I–V curve can be divided into three different segments, i.e., I, II, and III corresponding to the conductance values 0.3×10^{-10}, 6.99×10^{-10}, 1.0×10^{-10} S, respectively; this data further confirms the tunneling behaviour. The tunneling behaviour may be due to the contact potential between the nanowires and the probe; this (the contact potential) being a function of voltage (Mehrez and Guo 2004). Similar non-linear I–V behaviour has been observed for silver (Mehrez and Guo 2004), and gold (Wildoer et al. 1998) nanowires; such a behavior is required for functional electronic devices (Itakura et al. 1999). Figure 8b shows a quasi-plateau formation in region II, also observed by Cao et al. (2006), and the sharp increase and decrease in current noticed for the forward as well as the reverse bias correspond, respectively, to around +8.8 V and −9.4 V. For the three segments, the voltage, current, resistance and conductance are shown in Table 1.

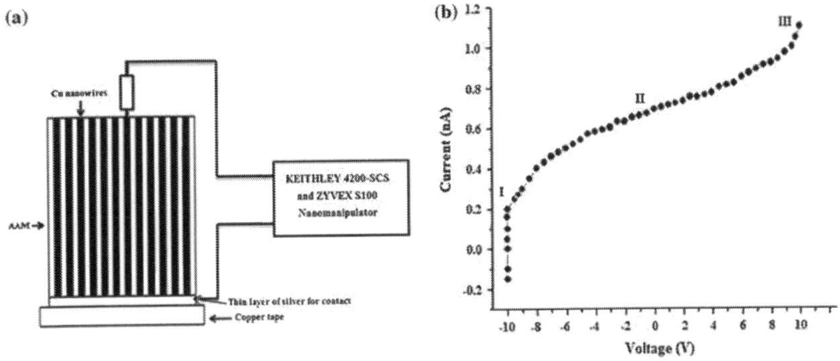

Figure 8. **a** The schematic of collective I–V characteristics measurement set up and **b** I–V characteristics curve of copper nanowires

Table 1. The measurement of voltage (V), current (I), resistance (R), and conductance, (S) corresponding to segments I, II, III of I–V characteristics

S. no	Region	Voltage (V)	Current (nA)	Resistance (Ω)	Conductance (S)
1	I	−10	0.3	33.33×10^9	0.30×10^{-10}
2	II	−1.0	0.7	1.43×10^9	6.99×10^{-10}
3	III	10	1.0	10×10^9	1.0×10^{-10}

Impedance analysis

Figure 9 stands for the impedance spectroscopy of the nanowires, embedded in AAM as host, at frequencies 0.2–1.0 MHz. An increase in impedance has been observed after frequency, 1 kHz, confirming inductive effect, which may be due to the mutual induction among the copper nanowires.

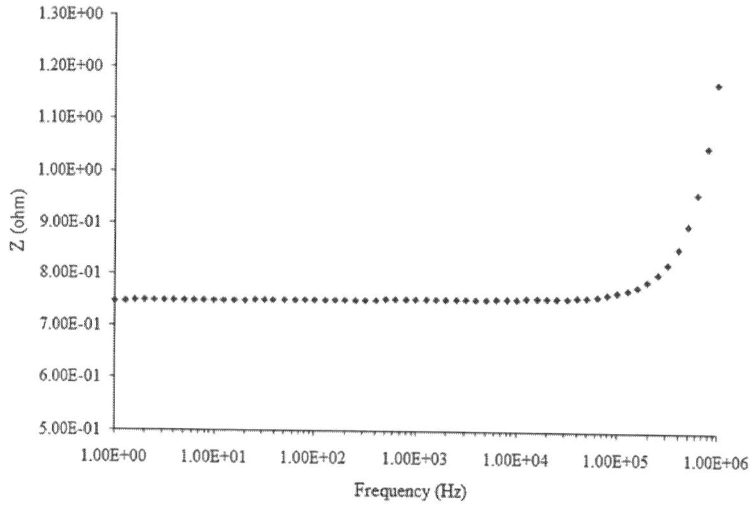

Figure 9. The impedance versus frequency behavior of copper nanowires

CONCLUSIONS

We have synthesized highly ordered and dense crop of copper nanowires using a template-assisted electrodeposition technique; the nanowires have been found to be of average size, 100 nm, and preferentially grown along the direction, [111]. The absorption spectra reveal that the peak, observed at 570 nm, is due to absorption of surface plasmon. The two excitation peaks, observed at 209 and 268 nm, yield a fluorescence peak at 296 nm in PL spectra, which is due to the fluorescence emission from the copper nanowires at different excitation wavelength..The non-linear I–V characteristics suggest formation of tunneling potential barrier in copper nanowires. The nanowires register rise in impedance above 1 kHz due to the inductive effect. The nonlinear I–V characteristics clearly show the possibility of the synthesized copper nanowires finding potential applications in future nanoelectronics. Moreover, these nanowires, being highly-ordered, may be directly patterned for fabrication of future nanodevices.

ACKNOWLEDGMENTS

Gurmeet Singh Lotey is extremely grateful to the Department of Science and Technology, India, for awarding him the INSPIRE fellowship to carry out the present research work.

REFERENCES

1. Bicelli LP, Bozzini B, Mele C, Durzo L (2008) A review of nanostructural aspects of metal electrodeposition. Int J Electrochem Sci 3:356

2. Boyd GT, Yu ZH, Shen YR (1986) Photoinduced luminescence from the noble metals and its enhancement on roughened surfaces. Phys Rev B 33:7923

3. Cao H, Wang L, Qiu Y, Zhang L (2006) Synthesis and I–V properties of aligned copper nanowires. Nanotechnology 17:1736

4. Choi H, Park SH (2004) Seedless growth of free-standing copper nanowires by chemical vapor deposition. J Am Chem Soc 126:6248

5. Duan J, Liu J, Mo D, Yao H, Maaz K, Chen Y, Sun Y, Hou M, Qu X, Zhang L, Chen Y (2010) Controlled crystallinity and crystallographic orientation of Cu nanowires fabricated in ion-track templates. Nanotechnology 21:365605

6. Furneaux RC, Rigby WR, Davidson AP (1989) The formation of controlled-porosity membranes from anodically oxidized aluminium. Nature 337:147

7. Gao T, Meng GW, Wang Y, Sun S, Zhang LD (2002) Electrochemical synthesis of copper nanowires. J Phys Condens Matt 14:355

8. Itakura K, Yuki K, Kurokawa S, Yasuda H, Sakai A (1999) Bias dependence of the conductance of Au nanocontacts. Phys Rev B 160:1163

9. Lai M, Riley DJ (2008) Templated electrosynthesis of nanomaterials and porous structures. J Coll Int Sci 323:203

10. Li AP, Muller F, Birner A, Nielsch K, Gosele U (1998) Hexagonal pore arrays with a 50–420 nm interpore distance formed by self-organization in anodic alumina. J Appl Phys 84:6023

11. Lisiecki I, Filankembo A, Kongehl HS, Weiss K, Pileni MP (1996) Control of the shape and the size of copper metallic particles. J Phys Chem 100:4160

12. Liu Z, Bando Y (2003) A novel method for preparing copper nanorods and nanowires. Adv Mater 15:303

13. Liu N, Wu D, Wu H, Liu C, Luo F (2008) A versatile and "green" electrochemical method for synthesis of copper and other transition metal oxide and hydroxide nanostructures. Mater Chem Phys 107:511

14. Mehrez H, Guo H (2004) In: Wang Zl (ed) Nanowires and nanotubes-materials, properties and devices-metal and semiconductor nanowires, vol 1. Springer, Berlin

15. Monson CF, Woolley AT (2003) DNA-templated construction of copper nanowires. Nano Lett 3:359

16. Mooradian A (1968) Photoluminescence of metals. Phys Rev Lett 22:185

17. Pan H, Sun H, Poh C, Feng Y, Lin J (2005a) Single-crystal growth of metallic nanowires with preferred orientation. Nanotechnology 16:1559

18. Pan H, Liu B, Yi J, Poh C, Lim S, Ding J, Feng Y, Huan CHA, Lin J (2005b) Growth of single-crystalline Ni and Co nanowires via electrochemical deposition and their magnetic properties. J Phys Chem B 109:3094

19. Qin Y, Staedler T, Jiang X (2007) Preparation of aligned Cu nanowires by room-temperature reduction of CuO nanowires in electron cyclotron resonance hydrogen plasma. Nanotechnology 18:35608

20. Schuchert IU, Toimil Molares ME, Dobrev D, Vetter J, Neumann R, Martin M (2003) Electrochemical copper deposition in etched ion track membranes. J Electrochem Soc C150:C189

21. Siwach OP, Sen P (2008) Fluorescence properties of Ag nanoparticles in water, spectrochimca acta part A. Mol Biomol Spectrosc 69:659

22. Toma A, Chiappe D, Boragno C, Mangeot FB (2010) Self-organized ion-beam synthesis of nanowires with broadband plasmonic functionality. Phys Rev B 81:165436

23. Wang C, Hu YJ, Lieber CM, Sun SH (2008) Facile synthesis of ultrathin and single-crystalline Au nanowires. J Am Chem Soc 130:8902

24. Wildoer JWG, Venema LC, Rinzler AG, Smalley RE, Dekker C (1998) Electronic structure of atomically resolved carbon nanotubes. Nature 391:59

25. Zach MP, Ng KH, Penner RM (2000) Molybdenum nanowires by electrodeposition. Science 290:2120

26. Zhang F, Wong SS (2009) Controlled synthesis of semiconducting metal sulfide nanowires. Chem Mater 21:4541

27. Zhang J, Sun J, Wei W, Shi S, Sun H, Guo J (2005) Synthesis of copper nanowires under a direct current electric field. Nanotechnology 16:2030

28. Zhang DW, Chen CH, Zhang J, Ren F (2008) Fabrication of nanosized metallic copper by electrochemical milling process. J Mater Sci 43:1492

INDEX